HC440.D5 C54 200

Chakravorty, San

Made in India :
 economic geogra
    2007.

2009 02 18

Humber College Library
3199 Lakeshore Blvd. West
Toronto, ON   M8V 1K8

MADE IN INDIA

To our fathers
Basudeb Chakravarty
and
Vinaykumar D. Lall
Our best teachers

MADE IN INDIA

*The Economic Geography and
Political Economy of Industrialization*

SANJOY CHAKRAVORTY
and
SOMIK V. LALL

OXFORD
UNIVERSITY PRESS

OXFORD
UNIVERSITY PRESS

YMCA Library Building, Jai Singh Road, New Delhi 110 001

Oxford University Press is a department of the University of Oxford. It furthers the
University's objective of excellence in research, scholarship, and education
by publishing worldwide in

Oxford   New York

Auckland Cape Town Dar es Salaam Hong Kong Karachi Kuala Lumpur
Madrid Melbourne Mexico City Nairobi New Delhi Shanghai Taipei Toronto

With offices in
Argentina Austria Brazil Chile Czech Republic France Greece Guatemala
Hungary Italy Japan Poland Portugal Singapore South Korea Switzerland
Thailand Turkey Ukraine Vietnam

Oxford is a registered trademark of Oxford University Press
in the UK and in certain other countries

Published in India by Oxford University Press, New Delhi

© Oxford University Press 2007

The moral rights of the author have been asserted
Database right Oxford University Press (maker)

First published 2007

All rights reserved. No part of this publication may be reproduced,
or transmitted in any form or by any means, electronic or mechanical,
including photocopying, recording or by any information storage and
retrieval system, without permission in writing from Oxford University Press.
Enquiries concerning reproduction outside the scope of the above should be
sent to the Rights Department, Oxford University Press, at the address above

You must not circulate this book in any other binding or cover
and you must impose this same condition on any acquirer

ISBN 13: 978-0-19-568672-2
ISBN 10: 0-19-568672-1

Typeset in ACaslon 10.5/12.5
by Sai Graphic Design, New Delhi 110055
Printed at Deunique, New Delhi-110 018
Published by Oxford University Press
YMCA Library Building, Jai Singh Road, New Delhi 110 001

Contents

List of Tables and Figures   vi

Acknowledgements   ix

1. Markets, States, and Industrialization   1
2. Patterns of Industrial Investment, Old and New   28
3. Determinants of Industrial Location   75
4. Economic Geography and the Firm   106
5. Industrial Clusters within Metropolitan Regions   142
6. On Spatial Policy   187

References   214

Index   230

Tables and Figures

TABLES

2.1	Long-term trends in industrial formation, 1961–94	31
2.2	Income and infrastructure indicators, 1950–89	32
2.3	Industrial classification codes and aggregation methods	41
2.4	Industrial investment by type of ownership, 1973–95	45
2.5	Statewise distribution of investment in selected sectors, pre- and post-reform	46
2.6	Top 25 districts and their shares, pre- and post-reform	47
2.7	Measures of industrial concentration and clustering, pre- and post-reform	56
3.1	Logistic models of new investment	86
3.2	OLS and Heckman selection models of new investment	88
3.3	Summary investment statistics by location	94
3.4	Determinants of probability of receiving private sector investment	95
3.5	Determinants of probability of receiving central government investment	96
3.6	Determinants of quantity of private sector investment	97
3.7	Determinants of quantity of central government investment	98
4.1	Characteristics of firms in the study sectors	119
4.2	Concentration in industrial sectors	120
4.3	Location and productivity	120
4.4	Number of establishments	125
4.5	Cost elasticities of production factors	126
4.6	Cost elasticities of economic geography variables	127
4.7	Input demand substitution	131
5.1	Indices of global clustering	156

5.2	Concentration in clusters	157
5.3	Correlation coefficients for industry pairs	176
5.4	Indices of co-clustering in selected industry pairs	177
5.5	Industry concentration in top districts	179
6.1	Examples of regional incentives	199

FIGURES

1.1	Transport costs and concentration of economic activity	8
2.1	Regions, states, districts, and metropolises	29
2.2	Investment distribution by region	49
2.3	Investment distribution by state	49
2.4	Investment distribution by metropolitan, urban, and coastal attributes	50
2.5	Investment distribution by city and suburb	50
2.6	Investment distribution by capital source	51
2.7	Distribution and clustering of total investment on a per-capita basis	57
2.8	Distribution and clustering of investment as share of national total	58
2.9	Distribution of investment by capital source	59
2.10	Clustering of investment by capital source	60
2.11	Distribution and clustering of heavy industry	61
2.12	Distribution and clustering of industry in the chemicals sector	62
2.13	Distribution and clustering of industry in the agribusiness sector	63
2.14	Distribution and clustering of industry in the textiles sector	64
2.15	Distribution and clustering of industry in the utilities sector	65
4.1	Maps of market access	116
4.2	Examples of industry concentration at the 3-digit level	121
5.1	Greater Bombay and its sub-regions	152
5.2	Metropolitan Calcutta and its sub-regions	153
5.3	Metropolitan Chennai and its sub-regions	154
5.4.1	Industry clusters in Greater Bombay, part 1	159
5.4.2	Industry clusters in Greater Bombay, part 2	160

5.4.3	Industry clusters in Greater Bombay, part 3	161
5.4.4	Industry clusters in Greater Bombay, part 4	162
5.5.1	Industry clusters in Calcutta metropolis, part 1	163
5.5.2	Industry clusters in Calcutta metropolis, part 2	164
5.5.3	Industry clusters in Calcutta metropolis, part 3	165
5.5.4	Industry clusters in Calcutta metropolis, part 4	166
5.6.1	Industry clusters in Chennai metropolis, part 1	167
5.6.2	Industry clusters in Chennai metropolis, part 2	168
5.6.3	Industry clusters in Chennai metropolis, part 3	169
5.6.4	Industry clusters in Chennai metropolis, part 4	170

Acknowledgements

This book would not have been possible without the support of several institutions and individuals. We would like to take this opportunity to recognize and thank them. Let us begin with institutions. The early stages of the project was funded by the US National Science Foundation's Geography and Regional Science programme (grant number SBR 9618343) and by a study leave granted by Temple University to Chakravorty. Later, Lall received funding from the World Bank (Research Grant 77960) that allowed us to continue to travel, acquire data, and employ research assistants. The Central Statistical Organisation (CSO) of the Government of India provided much of the data that are analysed in the following pages. We would like to thank the many individuals at CSO who were willing to be persuaded to release data that had long been cloistered.

We are indebted to several journals and their publishers for allowing us to use material that, in some form, had originally appeared in their pages. Specifically, we wish to thank: (1) Clark University for S. Chakravorty, 2000, 'How Does Structural Reform Affect Regional Development? Resolving Contradictory Theory with Evidence from India', *Economic Geography* 76: 367–94 (used in Chapter 2); (2) Taylor and Francis (www.tandf.co.uk) for S. Chakravorty, 2003, 'Industrial Location in Post-Reform India: Patterns of Interregional Divergence and Intraregional Convergence', *Journal of Development Studies* 40: 120–52 (used in Chapter 2); (3) John Wiley and Sons for S. Chakravorty, 2003, 'Capital Source and the Location of Industrial Investment: A Tale of Divergence from Post-Reform India', *Journal of International Development* 15: 365–83 (used in Chapter 3); (4) Blackwell Publishing for S. Lall and S. Chakravorty, 2005, 'Industrial Location and Spatial Inequality: Theory and Evidence from India', *Review of Development Economics* 9: 47–68 (used in Chapters 3 and

4); (5) Pion Limited, London for S. Chakravorty, J. Koo, and S. Lall, 2005, 'Do Localization Economies Matter in Cluster Formation? Questioning the Conventional Wisdom with Data from Indian Metropolises', *Environment and Planning A* 37: 331–53 (used in Chapter 5). The suggestions and critiques of the many anonymous reviewers provided by these journals and Oxford University Press have made the analyses more rigorous and the material more readable.

Superb research assistance was provided by several individuals. Foremost among them is Jun Koo, then at the World Bank, who worked through large quantities of data for Chapters 4 and 5. Bruce Boucek, then at Temple University, was very helpful in creating the early databases. Scholars at many institutions heard or read pieces of the book and offered useful suggestions and comments. These venues include: Jadavpur University and the Centre for Study of Social Sciences in Calcutta; the National Institute for Public Finance and Policy in Delhi; the World Institute of Development Economics Research in Tokyo (specially Tony Venables and Ravi Kanbur); conferences of the Regional Science Association, the Association of American Geographers; Temple University in Philadelphia, the University of California at Berkeley, the World Bank, etc.

Finally, we would like to thank our partners: Pallabi Chakravorty and Aditi Lall. They endured with grace and good humour the many hours that we spent hunched over keyboards, communicating with ape-like grunts rather than intelligible words. We are grateful for their unstinting support.

1

Markets, States, and Industrialization

Let us begin with a short journey. We start from the Hyatt Regency, a five-star hotel in Salt Lake, a suburb of Calcutta where many of the gleaming new offices of information technology giants like IBM, TCS, PWC, and Cognizant are located. We drive through the old congested and chaotic city, leaving behind a cluster of printing factories sitting cheek by jowl with a few city blocks populated by wholesalers and retailers of books, through a wholesale commodity market, across the river Hughli to Haora station. We catch a train to Bolpur in Birbhum district, about 150 kilometres away. Soon we are rushing past small stations set up to serve the iron and steel, jute, and cement factories dotting the banks of the river. After about 20 kilometres the factories begin to thin out and eventually disappear, and we are now in rural south Bengal, green with paddy fields and coconut groves. We see mud huts with roofs thatched with coconut fronds and doors so low that even the small undernourished women of the villages have to bend at the waist to pass through. By the time we reach Bolpur, the earth has turned reddish brown; it does not look capable of producing any crop. The town itself is dusty and small. There are no factories, only shops. It is a market town. We take a bus into the heart of the district where the Santhals, a minority group that is classified by the Government of India as a Scheduled Tribe, live in abject poverty. Their huts are meagre, their tools few and antiquated, the majority of the women are officially illiterate, many of the men effectively so. Their monthly earnings would not get them one room for one night in the place where we started, at the Hyatt.

There is not much that is surprising about this journey. It could be repeated in countless other cities and regions of the developing world. One could start in Mumbai or Bangkok, in north India or northeast Brazil. The starting points may differ in terms of average income. But such sub-national disparities in economic performance and living standards within and between regions are common, large, sustained, and, often growing. With empirical regularity, we observe high degrees of spatial concentration where a few cities account for much of the national employment and investment.

This is evident if we look at a map of Brazil, or Mexico, or Indonesia, or China, or India, using broad regional definitions (that is, at the state level) or using more fine-grained geographical demarcations (that is, at the district level). This uneven spatial distribution, whereby high income and productivity are concentrated mainly in large agglomerations, translates into differentials in economic performance across cities and regions within a country. Naturally, it raises questions about the growth potential for secondary cities, particularly those in lagging regions, and villages that are far less prosperous than even the secondary cities. This issue is especially important for developing countries as they have relatively lower levels of overall investment. Thus, regions that do not attract dynamic industries are not only characterized by low productivity, but also by lower relative incomes, standards of living, and development indicators such as rates of infant mortality, female literacy, and longevity.

Now, these indicators of development are also predictors of future growth. More educated populations are also more healthy and productive populations. In other words, there is a circular and cumulative relationship between current development and future development. Therefore, these large, persistent, and often growing differences in development standards across sub-national regions raise serious concerns on the development potential of lagging regions and a need to understand better the critical factors that influence the spatial distribution of economic activity within countries. While there has been much academic and policy interest in examining convergence or divergence of inter-regional economic performance over the past two decades, the findings from this body of work are contradictory, narrowly analysed, and usually do not provide, in the same pages, satisfactory answers on why we observe persistent economic disparities between geographic units and what role is played by policy and political economy to deepen or mitigate these economic differences?

WHY MARKETS AND STATES MATTER

We begin by making an obvious point: the story of modern economic growth is a story of industrialization. Economic growth is a feature of cities and regions which have industrialized, just as the absence of economic growth is a feature of cities and regions which have not industrialized. Notwithstanding isolated cases of cities that are entirely based on a local resource such as a natural attraction or a specialized agricultural product, in general, there is little doubt that the modern city is an industrial city. Even post-industrial growth, which is characteristic of the more developed nations today, is based on the foundations created by industrial growth. These differential growth patterns between more and less industrialized regions have created widely differentiated sub-national spaces leading to significant spatial inequalities at different scales. For instance, in India, the average income ratio of the richest state to the poorest has grown from about under 2.5 to over 3.5 within fifty years of independence.

However, success at industrialization or the failure to industrialize is not determined entirely in the local region in question. Success, failure, and intermediate outcomes are the result of some factors of economic geography such as proximity to resources and markets, transportation costs, localization and urbanization economies, etc. (that is, factors that are market-based) and some factors of political economy such as historical path dependence, policy decisions on infrastructure, exchange rates, land use, globalization, etc. (that is, factors that are strongly influenced by the state). Hence, the fact that the extent and type of industrialization varies over space is best understood and explained in terms of economic geography and political economy. The forces of 'pure' economic geography are market forces, but they are mediated by political economy, or the state. Therefore, industrialization processes have to be understood in terms of the interaction of markets and states.

This is the approach we bring to this analysis of industrialization in India. From independence in 1947, the Indian state has emphasized the role of manufacturing industry in 'catching up' to the industrial economies of the west. Today, India is a major industrial nation. Its output includes aircrafts, ships, cars, locomotives, heavy electrical machinery, construction equipment, power generation and transmission equipment, chemicals and petrochemicals, pharmaceuticals, precision instruments, communication equipment, and computers. Its industrial output has grown by over 7 per cent per year over the preceding two decades, its Gross Domestic Product (GDP) has grown by 6.8 per cent per year since 1994, and its economy (in terms of purchasing power parity) is estimated to be the fourth largest in the world.[1] The term 'Made in India' does not yet have global cachet, but for the large domestic economy of one billion consumers, it is an ubiquitous emblem.

This book then is the definitive account of the geography of industrialization in India. Our objectives are threefold:

(1) To describe and analyse the processes of industrial formation over the long run (especially over the last decade and half), at multiple spatial scales (region, state, metropolis, district, and pin code), and the concomitant rising tide of spatial inequality.
(2) To show how market forces (economic geography) and state actions (political economy) have contributed to this condition.
(3) To interrogate some of the most fundamental questions in economic geography—specifically the issues surrounding market access and external economies—and make significant contributions to its theoretical foundations.

To start with, it is necessary to resolve several contradictions in theory and policy. For instance, in the economic literature, typical regional convergence analyses (Barro and Sala-i-Martin 1992, 1995 and followers) are built on the assumption of *decreasing returns to reproducible factors*, where income disparities arising from differences in regional capital/labour ratios diminish over time. Thus, these models predict *convergence* between regions as both trade and factor flows tend to equalize factor prices. However, there is another large literature which argues that increasing returns (often seen as arising from technological or pecuniary agglomeration externalities) are more likely than decreasing returns. These generate a process of circular and cumulative causation leading to decreasing costs of production and continuing concentration. Coupled with the fact that there are significant barriers to labour mobility, these cumulative processes result in conditions in which a significant number of people continue to live in lagging regions. We do not deny the existence of forces that may lead to inter-regional convergence in some cases, but we are persuaded that increasing returns and cumulative and circular processes are dominant for very long periods. Throughout this book, in different ways, we keep returning to this conclusion.

What does the state do in response to these large and persistent regional inequalities? Most states have opted for interventions to offset some of these market pressures and to promote relatively balanced regional development. This has created a policy tension or contradiction between the market solution of migration or labour flows, that is, 'moving people to jobs', and the interventionist solution of 'moving jobs to people' or promoting capital flows (including a variety of fiscal transfers whether to subsidize credit creation, job creation, employment of local people, or income support) as well as by providing public goods in lagging regions. These policy contradictions are quite apparent. Less obvious are the contradictions between what have been called 'explicit' spatial policies (designed to favour lagging regions) and 'implicit' spatial policies (on exchange rates, import substitution, energy prices, land use, and other arenas) whose impact often overwhelms the influence of explicit policies. Add to these contradictions the dilemma faced by developing nation states in trying to respond to globalization: cities, especially big cities, are the places where products can be made for the global marketplace. Hence, these cities have to be promoted and upgraded, which is typically done at the cost of promoting lagging regions. In short, there are tensions between spatial efficiency and equity.

State policies have an impact that perpetuates the processes of divergence for very long periods. Cumulative causation works in two ways—upward and downward—creating spirals of growth in some regions and either decline or stagnation in others. These are processes that operate over a very long run, partly because they set into motion actions by individuals, groups, and sub-national states which often tend to perpetuate the divergent growth paths. For instance, migrants with human and financial capital leave lagging regions for advanced ones, leaving the poor region poorer and contributing to the growth of the relatively rich region. One of the results is that, growing regions pay more attention to growth, whereas lagging regions focus on the politics of identity and class. It is possible to show, for instance, that the method of revenue collection used by the British colonizers has strongly influenced the terms of the discourse on identity and class in specific regions; this, in turn, has influenced the political economy of regions in the post-independence period.

It is in this context of contradictions (in political economy and economic geography) that we set our examination of industrialization in India. In the remainder of this chapter we spell out the principal theoretical issues which concern us. These issues are relevant not only in India, but for spatial development anywhere. Our review of the theoretical issues is organized as follows:

(1) First, we turn to the micro-foundations of economic geography and review the main findings and predictions from recent analytic and empirical work in the 'new economic geography' literature as well as the more traditional regional science literature to identify factors that influence the location and growth of economic activity across regions.
(2) Next, we review the principal strands of literature on the political economy of regional development by examining the relationship between markets and states as manifested in spatial terms.

SPATIAL CONCENTRATION OF ECONOMIC ACTIVITY

In this section, we review the relevant literature on spatial concentration and growth of economic activity. This review is by no means exhaustive but is aimed at identifying the key issues that emerge from existing analytic and empirical work. Research on location and concentration of economic activity has long been of interest to economists, geographers, planners, and regional scientists (Greenhut and Greenhut 1975; Hotelling 1929; Isard 1956; Lösch 1956; von Thunen 1966; Weber 1929). However, analytic difficulties in modelling increasing returns to scale marginalized the analysis of geographic aspects in mainstream economic analysis (Krugman 1991a). Recent research on externalities, increasing returns to scale, and imperfect spatial competition (Dixit and Stiglitz 1977; Fujita *et al.* 1999; Krugman 1991b) has led to a renewed interest in analysing the spatial organization of economic activity. This is especially true in the case of geographic concentration or clustering.

Models in the 'New Economic Geography' (NEG) literature (see review in Fujita, Krugman, and Venables 1999) allow us to move from the question 'Where will industry concentrate (if it does)?' to the question 'What industry will concentrate where?' These insightful theoretical models provide, for the most part, renewed analytical support for the 'cumulative causation' arguments made in earlier decades on the core–periphery relationship, on agglomeration economies, and on industrial clustering. The main findings from the economic geography literature can be organized in two categories:

(1) Market access and transport costs
(2) Agglomeration economies

We review both these factors in this section.

MARKET ACCESS AND THE COSTS OF REMOTENESS

In traditional location models, *production is assumed to take place under conditions of constant or diminishing returns to scale.* Under these conditions, firm location decisions are based on the fact that transportation has costs associated with it. One implication is that industry is likely to spread out to minimize the costs of reaching consumers in different parts of the country. The 'folk theorem' of spatial economics (Fujita and Thisse 1996) says *that under conditions of constant or diminishing returns to scale there will be very many small plants supplying local markets.* However, in the presence of increasing returns to scale, firms are able to concentrate production in relatively few locations, and make choices on where to operate (Henderson, Shalizi, and Venables 2001). These models of location choice with increasing returns and imperfectly competitive market structures are developed in the NEG literature.

Krugman's (1991b) seminal paper shows that increasing returns activities are pulled disproportionately towards locations with good market access. For example, if there are nine locations, in eight of which the share of final expenditure is 10 per cent, and one for which that share is 20 per cent, then other things being equal, more than 20 per cent of manufacturing supply will be met from this larger location. The reason is simply the benefit of having low transport cost access to this large market in comparison to more expensive access to other markets. This immediately creates a force for the agglomeration of activity. As a disproportionate share of manufacturing is attracted to a location, either the wage rate in the location will increase or labour will be attracted to immigrate—either of which will tend to increase this location's share of total expenditure still further. The market access effect is sometimes called the 'home market effect', and this combined with labour mobility is the basis of Krugman's thesis.

The extent to which market access enters into the location decision, depends on the level of transport costs. If transport costs are very high, then activity is dispersed. In the extreme case, under autarky, every location must have its own industry to meet final demand. On the other hand, if transport costs are negligible, firms may be randomly distributed, as proximity to markets or intermediate suppliers will not matter. It is only at intermediate levels of transport costs that agglomeration would occur, especially when the spatial mobility of labour is low (Fujita and Thisse 1996). We therefore expect a bell shaped (or inverted U-shaped) relationship between the extent of spatial concentration and transport costs (see Figure 1.1).

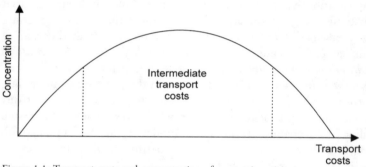

Figure 1.1: Transport costs and concentration of economic activity

Source: Adapted from Fujita, Krugman, and Venables (1999).

In principle, improved access to consumer markets and intermediate buyers and suppliers will increase the demand for a firm's products, thereby providing the incentive to increase scale and invest in cost-reducing technologies. With a decline in transport costs, firms have an incentive to concentrate production in a few locations to reduce fixed costs. Transport costs can be reduced by locating in areas with good access to input and output markets. Thus, access to markets is a strong driver of agglomeration towards locations where transport costs are low enough that it is relatively cheap to supply markets due to availability of quality transport networks (Henderson *et al.* 2001). In addition to the pure benefits of minimizing transport costs, the availability of high quality infrastructure, linking firms to urban market centres increases the probability of technology diffusion through interaction and knowledge spillovers between firms, as well as between firms and research centres, and also increases the potential for input diversity (Lall, Shalizi, and Deichmann 2004). As a result, improved accessibility has the effect of reducing geographic barriers to interaction, which increases specialized labour supply and facilitates information exchange, technology diffusion, and other beneficial spillovers that have a self-reinforcing effect.

AGGLOMERATION ECONOMIES—LOCALIZATION

In addition to market access, firms tend to concentrate production to benefit from localization economies, which are externalities that enhance productivity of all firms in that industry. At the industry level, scale economies accrue to firms due to the size of the industry in a particular location. These economies are external to the firm but internal to the industry. There is considerable theorizing on localization economies in the works of Alfred Marshall (1890), Kenneth Arrow (1962), and Paul Romer (1986); these are often called MAR externalities (from the initials of the primary contributors). They argue that cost-saving externalities are maximized when a local industry is specialized, and their models predict that externalities predominantly occur within the same industry. Therefore, if an industry is subject to MAR externalities, firms are likely to locate in a few cities where producers of that industry are already concentrated. Examples of highly localized industries are ubiquitous. Semiconductor and software in Silicon Valley and automobile in Detroit are classic cases in point. In India, we have had textiles in Mumbai and jute in Calcutta. Later, Michael Porter (1990) emphasized the importance of dynamic externalities created in specialized and geographically concentrated industries.

Benefits from localization include sharing of sector-specific skilled labour, sharing of tacit and codified knowledge, intra-industry linkages, and opportunities for efficient subcontracting. Further, the presence of a disproportionately high concentration of firms within the same industry increases the possibilities for collective action to lobby regulators or bid down prices of intermediate products (Lall, Shalizi, and Deichmann 2004). These location-based externalities imply that firms are likely to benefit from locating near large concentrations of other firms in their own industry. In addition to the supply-side linkages discussed above, localization economies are also realized on the demand side. These include reduction of information asymmetries for consumers as well as the ability to attract price and quality comparison shoppers. The existence of auto malls, jewellers rows, bookstore and restaurant enclaves in urban areas are examples of this phenomenon. These so-called 'thick-market externalities' benefit all firms in an industry located in close geographic proximity and can occur in relative isolation from other industries. There is an extensive empirical literature supporting the positive effects of localization economies on economic performance (Henderson 1988; Ciccone and Hall 1995). The benefits of own-industry concentration can, however, be offset by costs such as increased competition between firms for labour and land causing wages and rents to rise, as well as increased transport costs due to congestion effects. Firms in industry sectors which predominantly use standardized technologies and low skilled workers for production may not benefit enough from intra-industry externalities to offset costs from increased own-industry concentration.

In a study of Korean industry, Henderson *et al.* (2001) estimate scale economies using city level industry data for 1983, 1989, and 1991–3, and find localization economies of about 6 to 8 per cent. Lall and Mengistae (2005) use firm survey data from India and find that own-industry concentration has a significant bearing on firm location decisions across cities, and this effect is the highest for technology-intensive sectors. In another recent study using firm-level data from Indonesia, Deichmann *et al.* (2005) also find evidence that localization economies are higher for high technology (office computing) and natural resource-based industries (wood and rubber and plastic) and lower for footloose industries such as garments and textiles. Their policy experiments suggest that in the presence of agglomeration economies, increases in infrastructure endowment for lagging regions may have only limited pay-offs in terms of attracting firms from other more established leading regions, particularly in mainstream sectors that have already concentrated in other leading regions.

Agglomeration Economies—Inter-industry Linkages

The third force (following market access and localization economies) comes from combining own-industry concentration with the production of intermediate goods. Demand for manufacturing comes not just from final consumers but also from intermediate demand or inter-industry linkages. Therefore, a location with a high share of firms will have a high demand for intermediates, which further increases its attractiveness for manufacturing firms. In addition to these demand effects there are cost benefits; as a large number of intermediate suppliers are attracted into the location, firms using intermediate goods can save on transport costs, making the location still more attractive.

The importance of inter-industry linkages as a major agglomerative force was first recognized by Marshall (1890, 1919). Venables (1996) demonstrated that agglomeration could occur through the combination of firm-location decisions and buyer–supplier linkages even without high factor mobility. The presence of local suppliers can reduce transaction costs and therefore increase productivity. Inter-industry linkages can also serve as channels for vital information transfers. Firms that are linked through stable buyer–supplier chains often exchange ideas on how to improve the quality of their products or on how to save production costs. It is such ongoing interactions that make the dynamics of inter-industry externalities so vibrant. Therefore, if the performance of an industry is highly dependent upon the supply of high quality intermediate goods (for example in automobile manufacturing), firms are likely to locate in regions with a strong presence of local suppliers. The presence of local supplier linkages makes buyer industries more efficient and reinforces the localization process.

The empirical evidence from developing countries on the importance of intermediate suppliers or inter-industry linkages in influencing location decisions and industrial performance is still in its infancy. Recent work by Amiti and Cameron (2004) shows that externalities which arise from inter-industry linkages are highly localized and have a significant impact on manufacturing performance (measured by wages) in Indonesia. Location models in Deichmann et al. (2005) also find that access to suppliers influences location decisions of firms in several industry sectors (food and beverages, garments, chemicals, rubber). However, firm-profit models estimated in Lall, Funderburg, and Yepes (2004) for Brazil do not find significant gains from supplier access, when they control for market access and other sources of agglomeration economies. For China, Amiti and Javorcik (2005) find that market and supplier access in the province of entry are the most important factors affecting foreign entry, which is consistent with market fragmentation due to underdeveloped transport infrastructure and informal trade barriers.

AGGLOMERATION ECONOMIES—URBANIZATION ECONOMIES

Scale economies from urbanization emanate from the overall size (not only in terms of the number of firms but also in terms of population, income, output, or wealth) and diversity of the urban agglomeration. For a firm, benefits from urbanization include access to specialized financial and professional services, inter-industry information transfers, and availability of general infrastructure such as telecommunications and transportation hubs. Size is usually correlated with diversity, as larger urban areas can support a wider range of activities. Small cities are specialized in a few manufacturing activities, or are either administrative centres (such as regional capitals or university towns in some countries), or agricultural market centres providing services for farmers. In comparison, larger cities are more diverse, supporting a variety of manufacturing activities that require buyers and suppliers to be in close spatial proximity (input–output linkages). Further, larger cities are centres of innovative technologies and usually tend to offer business or productive services.

The importance of urbanization economies arising from industrial diversity is linked to the work of Chinitz (1961) and Jacobs (1969). In their representation, diversity provides a summary measure of urbanization economies, which accrue across industry sectors and provide benefits to all firms in the agglomeration. They propose that important knowledge transfers primarily occur across industries and the diversity of the local industry mix is important for these externality benefits. They argue that cities are breeding grounds for new ideas and innovations due to the diversity of knowledge sources concentrated and shared in cities. The diversity of cities facilitates innovative experiments with an array of processes, and therefore new products are more likely to be developed in diversified cities. Therefore, industries with Jacobs type externalities tend to cluster in more diverse and larger metro areas.

The benefits of locating in a large diverse area go beyond pure technology spillovers. Firms in large cities have relatively better access to business services, such as banking, advertising, and legal services. Particularly important in the diversity argument, is the heterogeneity of economic activity. On the consumption side, the utility level of consumers is enhanced by increasing the range of goods that are available locally. At the same time, on the production side, the output variety in the local economy can affect the level of output (Abdel-Rehman 1988; Fujita 1988; Rivera-Batiz 1988). That is, urban diversity can yield external scale economies through the variety of consumer and producer goods. Recent empirical studies by Bostic (1997) and Garcia-Mila and McGuire (1993) show that diversity in economic activity has considerable bearing on the levels of regional economic growth.

There is considerable empirical work which examines the contribution of urbanization economies on productivity. In one of the earliest studies, Sveikauskas (1975) used manufacturing data for the US at the two-digit SIC (Standard Industrial Classification) level and found that a doubling of city size increased labour productivity by 6 per cent. Using Japanese data, Tabuchi (1986) found that a doubling of population density increases labour productivity by 4.3 per cent. The results from empirical studies on the relative importance of specialization and diversity are mixed. Glaeser *et al.* (1992) find evidence only in favor of diversity. On the other hand, Miracky (1995) finds little evidence to support the diversity argument. For Indonesia, Henderson *et al.* (1995) show that the significance of diversity is different for different industrial sectors. They find evidence of specialization externalities in mature capital goods industries and of diversity externalities in new high-tech industries. These findings are consistent with the product cycle theory (Vernon 1966), which predicts that new industries tend to prosper in large and diverse urban areas, but with maturity, their production facilities move to smaller and more specialized cities.

SUMMARY

In the final analysis, localization economies, input–output linkages, and urbanization economies are not mutually exclusive. They may occur individually or in combination. For instance, consider whether it is possible to have localization economies without urbanization economies? Imagine a production centre which is populated by fifty small and large firms that produce only one item, say cricket bats. The region produces willow of a type that is well known for being turned into high quality cricket bats. These fifty firms employ on average sixty employees, so the number of factory workers is 3000. If we assume that each factory job has a multiplier effect of 2.5 in the formal sector (that is, for each factory job there are tertiary jobs in finance, education, health care, retail, etc., which total to 1.5 jobs) and 2.0 in the informal sector (transportation workers, domestic servants, etc.), then the total number of workers in this production centre is 10,500. Let us also assume that each worker supports four other individuals. Then we can have a town of 52,500 people that is based on a single product. This centre is too small to realize general urbanization economies. Hence, we can technically argue that it is possible to realize localization economies without urbanization economies. However, such instances are rare and usually based on a local resource (willow in this case; merino wool and champagne are other examples). In most cases several spatial economies operate together and part of the difficulty faced by analysts of such forces is that it is not easy to separate out the effects of each. One of our main tasks in this book is to separate these effects. We use different methods to do so, and in Chapters 4 and 5 we argue that urbanization economies and industrial diversity may be the most important of the external economies as far as industry location is concerned.

POLITICAL ECONOMY OF REGIONAL DEVELOPMENT

Do these market (or economic geography) forces operate without any state intervention? Of course not. The presence of the state is hinted at in the idealized account presented above. Who builds the roads and rail lines that form the transportation networks that are so critical in reducing costs? With rare exceptions, it is the state. Therefore, from the beginning of development practice and scholarship, questions on regional development have been framed around the institutions of the state and the market: What are the possible outcomes for comparative regional development under free market conditions (with unrestricted factor mobility)? What kinds of state intervention are possible and/or necessary for achieving balanced growth? What are the effects of state intervention on comparative regional development (the equity issue), and on the nation's development prospects (the efficiency issue)? This literature is very familiar to students of regional development. We divide this literature and the knowledge generated by it into two parts; the point of departure is the break created by liberalization and structural reforms in many nations. This break is also associated with globalization.

REGIONAL DEVELOPMENT THEORY—
THE PRE-GLOBALIZATION PHASE

Economic models of regional development are built on the general assumption that increasing returns and diminishing returns are both possible in the long run; and since the latter comes later, inter-regional convergence is the likely long-term outcome. These models tend towards equilibrium and convergence, rest on export-driven growth and the economies of agglomeration in dynamic nodal regions, where most regions derive long-term benefits from modernization and technical change (Borts and Stein 1964; Isard 1975; North 1975; Richardson 1973). However, there are contradictions between the economies of scale and agglomeration on the one hand, and size related congestion diseconomies on the other in metropolitan regions (Petrakos 1992; Wheaton and Shishido 1981). Or, in Krugman's (1991a, 1995) terms: there is tension between centripetal forces (higher labour productivity, larger plant size, access to markets and products, that is, backward and forward linkages, thick labour markets, and knowledge spillovers) and centrifugal forces (higher land rents, commuting costs, congestion and pollution, all leading to higher wages and taxes). For indeterminately long periods after industrial development begins, large cities offer increasing returns to capital investment. Eventually, because of lower transportation costs, the costs of size related congestion rise above the benefits of concentration-based externalities, so that higher returns become possible in smaller urban centres.

An early culmination of this approach was Jeffrey Williamson's (1965) famous thesis on the inverted-U curve of regional inequality. Williamson used arguments similar to the ones outlined by Kuznets (1955) to suggest that regional inequality increases during the early stages of development, and declines during the later stages. His cross-sectional analysis of regional inequality in countries at different levels of development appeared to provide solid empirical support for his hypothesis, and ever since, the 'Williamson hypothesis' has become one of the cornerstones of the regional development literature. Alonso (1980) went so far as to suggest that the Williamson curve is one of the five bell-shaped curves that almost invariably characterize the development process (see Chakravorty 1994 for a counter-argument).

By the early 1980s, however, conventional neoclassical growth theory (the Solow model) had, in the view of many scholars, become unrealistic in that it treated technological change and human capital as exogenous factors. As a result, a 'new' growth theory that treats these variables as endogenous to the growth process has gained legitimacy. After the pioneering, non-spatial work of Romer (1986) and Lucas (1988), it has been extended by Barro and Sala-i-Martin (1995), Armstrong (1995), Sala-i-Martin (1996) and others to the regional realm (where, it is presumed, the model assumptions are more reasonably satisfied). There are several competing and overlapping versions of such endogenous models, and as shown by Martin and Sunley (1998), their theoretical expectations, from a spatial or non-spatial perspective, are contradictory.

The models predicting convergence were not, however, the first perspectives on comparative regional development. Some pioneers in development economics—Gunnar Myrdal (1957) and Albert Hirschman (1958)—had been sceptical about the growth prospects of lagging regions. They suggested the *core–periphery* and *cumulative causation models* which Kaldor (1970), Friedmann (1966, 1973), Perroux (1950), Boudeville (1966) and others subsequently extended. In this view, because the natural forces of economic geography tend to favour the existing leading regions (the core), they would tend to grow faster than the existing lagging regions (the periphery). Therefore, if there is no state intervention, regional imbalances are likely to widen. Because of demands from the periphery, state intervention is not only politically necessary and inevitable; it also improves the distribution of welfare. Myrdal, however, was frankly pessimistic about the prospects of lagging regions. According to him:

It is easy to see how expansion in one locality has 'backwash effects' on other localities. More specifically the movements of labour, capital, goods, and services do not by themselves counteract the natural tendency to regional inequality. By themselves, migration, capital movements, and trade are rather the media through which the cumulative process evolves—upwards in the lucky regions and downwards in the unlucky ones. In general, if they have positive effects for the former, their effects on the latter are negative (Quoted in Higgins & Savoie 1995: 86).

A more hopeful position was taken by Friedmann and Hirschman—they saw the core as the locus of change, where new ideas, technology, and capital intersect to generate economic and cultural dynamism. Granted, they argued, that the non-metropolitan periphery initially falls behind, but eventually expanding markets and urbanization, the spatial diffusion of innovations and culture, and political demands from the periphery (mediated by state actions) should lead to some narrowing of the core–periphery gap.

Marxist and Neomarxist models of the political economy of regional development have been far more critical of the bourgeois state. Dependency theorists such as Paul Baran (1957), Andre Gunder Frank (1967), Michael Timberlake (1987), and others have argued that deepening class polarization and geographical inequality are the outcomes of modernization and industrialization. In this view, subnational uneven development is a less important subset of the fact of deepening uneven development between nations. The core–periphery system that exists at the intra-national scale also exists at the international scale. The periphery states are dependent on the core states for technology and demand for goods. The elite in these dependent-periphery states assist the capital owners from the developed core to extract surplus, and the underdevelopment of the regional and international periphery is a necessary condition for the development of the core.

In the more subtle views of Harvey (1982) and Massey (1984), regional change is episodic, where neither divergence nor convergence holds over the long run. More recently, de-industrialization or post-industrialization in developed nations has been theorized under the perspectives of 'post-Fordism' or 'flexible accumulation' (see Piore and Sabel 1984; Scott 1988; Storper ans Walker 1989). Here the emphasis is on transactions costs and 'the character of technological change, the form and organization of firms and industries, (and) the creation and transformation of labour markets' in influencing regional change (Schoenberger 1989: 133). These ideas have been extended to the context of third world development (Diniz 1994; Storper 1991), but there has been little systematic or system-wide extension of this approach. The followers of Michael Lipton and his 'urban bias' thesis (Lipton 1977) were just as pessimistic as the Neomarxists. In their view, state policies are designed to benefit urban areas at the cost of rural areas: to extract the rural surplus and invest it in more productive urban enterprise. As a result, lagging rural regions fall even further behind the growing urban regions.

Hence, the role of the state in influencing urban and regional development has been theorized from several perspectives—from 'urban bias' and 'dependency' theorists who have argued that policies are created with the intent to favour urban areas or metropolitan centres, to the view that state interference distorts market determined spatial distributions (of population and investment) in unintended and unforeseeable ways, or has little influence (Brooks 1987; Henderson 1988). The set of 'explicit' urban and regional policies aimed at industrial decentralization—licensing control, location of public sector projects, promotion of industrial estates, tax and land acquisition incentives, etc., in favour of lagging regions—has often been subsumed by more powerful macroeconomic policies (which may be called 'implicit' spatial policies) on development path, subsidies to industry relative to agriculture, monetary, exchange rate, and trade policies, etc. (Henderson 1982; Mills 1987). Despite the overall failure of regional policies (which are subsumed by implicit spatial policies), they doubtless have had some effect: perhaps one only of retardation rather than reversal of metropolitan polarization.

To summarize the confusion: regional theory suggests that regional differences are likely to widen in the absence of state intervention (Myrdal), which is not necessarily a negative outcome (Friedmann), as, in the long run regional differences will decline anyway (Williamson, Barro and Sala-i-Martin). State intervention is a necessary aspect of the political process (Hirschman), but is biased toward urban areas (Lipton) and developed nations (Baran, Frank), and may lead to inter-regional convergence (Sala-i-Martin; Krugman) or is inefficient or irrelevant or dwarfed by 'implicit' regional policies (Henderson; Richardson).

REGIONAL THEORY AND GLOBALIZATION

There is widespread agreement among analysts that economic globalization has ushered in a new phase in inter-regional development. Economic globalization is marked by increasing international trade, a more rapid and voluminous flow of goods, services, information, and investment across international boundaries and ever longer distances, and a new international division of labour, where routine manufacturing takes place in developing nations, and design and control functions are undertaken in more developed nations. Economic globalization is accompanied by, and largely made possible by, ideological globalization, whereby developing nations adopt similar sets of policies that enable trade (Chakravorty 2003). These policies have been variously termed: the Washington Consensus (Williamson 1990, 2000), liberalization, or structural reforms. We do not need, at this point, to understand the differences between these terms.

This shift to market-friendly liberalism over the last decade-and-a-half signifies an important paradigm shift. Economic and political nationalism was the idea that led to the creation and continuance of the nation-state in the post-colonial world. That idea appears to be losing ground to the twin forces of economic and ideological globalization. The role and nature of the state is also changing as a consequence. It is necessary to focus on the relationship between the new, reformed state and its policies, and the resulting development impact at the sub-national level. Historically, the state has been instrumental in shaping the economic geography of regions in the developing world—starting with the establishment and privileging of port cities for external trade and administration during the colonial period, to the creation of a complex array of rules and regulations that established location incentives and disincentives during the nationalist period.

The critical questions after reform are: what is the role of the state in shaping the economic landscape after the nature of the state has been altered to a (supposedly) less interventionist version; and, how do domestic and international actors respond to the freer markets, especially in terms of investment location? One assumption generally shared in the regional development literature we have outlined in the previous pages is that of policy continuity, if not regime continuity. It is apparent now that this assumption is invalid in most situations: liberalization or structural reform is a fundamental shift away from the policies of the past. As a result, it is now necessary to formulate a new theoretical framework for the analysis of regional development.

What are the general policy imperatives of the post-reform state? In the old nationalist model, the national state tried to be the principal agent of economic change using an institutional and regulatory structure that emphasized centralization over federalism, state ownership of heavy industry and infrastructure over private ownership, and self-reliance or import substitution over export orientation. In the new liberal model, the state is significantly less involved in the ownership of industry and the regulatory structure affecting new investments; there are lower entry barriers to multinational capital; export orientation is favoured over import substitution; and steps toward some decentralization of power and policy instruments in favour of sub-national states are taken.

But the 'new' state is far from a hands-off, free trading, *laissez faire*, purely market-enthusiastic entity. One of our basic premises is that the neo-liberal nation-state is bipolar in more ways than one—for instance, 'there is a coexistence of liberalizing and protectionist policies' (Leinbach 1995: 204)—which leads to inaction in some arenas while simultaneously there is more concerted action in other arenas (see Wade 1990 and Brohman 1995 on this thesis). As far as regional development is concerned, the newly liberal state is both a reduced or spatially disengaged state (as far as the promotion of regional balance is concerned), and a more enlarged state in terms of promoting selected metropolitan regions for receiving investment, especially foreign direct investment (FDI). These changes are concomitant with more active sub-national states competing in asymmetrical spatial structures shaped by colonialism, and subsequently nationalism, to capture new and different markets. The altered geography of investment opportunities suggests altered inter-regional development possibilities. The fact that structural reform is a discontinuity in the development process is, however, merely one of the complications: regional development theory, which was already beset with contradictions from different ideological perspectives (as shown in the foregoing discussion), has been further complicated with the regional implications of the increasingly dominant new or endogenous growth theory. Let us look at the issues from a pragmatic, political economy perspective.

What are the significant factors affecting regional change in developing nations after liberalizing reforms? Elizondo and Krugman (1992) suggest that post-reform regional development is likely to be more evenly balanced. They argue that the magnitude of internal trade is much larger than foreign trade in inward looking trade regimes; 'this leads to concentration of production and trading activities in large metropolitan cities... an opening up of the economy is likely to break the monopoly power of these highly concentrated production and trading centers, weaken the traditional forward and backward linkages and lead to a more even distribution of economic activities across regions' (Das and Barua 1996: 365). Similarly, according to Gilbert (1993: 729), 'the cities which benefited most from the previous development model have suddenly had an important prop to their growth removed' in the new model of liberalization and export orientation. These could be valid arguments, of course, but should be considered in the light of actual experience—much depends on the extent of urban and metropolitan bias before reforms, the size of the domestic market, the nature of the regional hierarchy, the degree of protectionism, the quantity of international trade, the specific liberalization strategies adopted, etc. (see Markusen 1995).

In developing nations like India, perhaps the most important structural factor to consider is the availability of infrastructure. Metropolitan regions have, by far, the highest standards of physical infrastructure (in power, roads, housing, telecommunication, etc.) and social infrastructure, such as schools and hospitals (Kessides 1993). During the stage of state controlled development, this infrastructure concentration factor was somewhat offset for two reasons. First, the state itself was the primary decision maker on where much of the capital was invested; earlier, significant proportions of this investment had gone to infrastructure and heavy industry near raw material sources (in what have often been called growth centres) in non-metropolitan locations. Often these centres became enclaves, linked to the national and international economy rather than the local economy, but industry was decentralized somewhat. Second, as indicated earlier, incentives in lagging regions and disincentives in metropolitan regions raised the former's share in total investment beyond what a no-policy situation would have allowed.

After market-oriented reform these two factors are drastically changed. The state moves away from industrial ownership, and therefore location decisions, by allowing private industry to participate in all industrial arenas, and by eventually divesting state-owned industry to the private sector. Egalitarian regional policies are withdrawn or not enforced. At the same time, following the East Asian model, the national state sees foreign investment as the key to spurring economic growth; it also logically sees its metropolises as likely foreign investment destinations. It invests in infrastructure in the leading metropolises, and encourages competition between cities and regions for other investments (Cook and Hulme 1988). Subnational governments, with more freedom to enact policies, tend to react by further eroding local environmental and egalitarian policies, and by emphasizing the growth prospects of the largest cities. The coastal regions in general, and the coastal cities in particular, become the focal points of directed investments, as in the new, outward-looking regime, access to the outside world from coastal locations assumes greater importance.

Finally, the state takes action to reconfigure the metropolis to accommodate the new growth. Many of the concerns with primacy and the advocacy of decentralization stem from the perception that the existing metropolitan cores are very congested, and, therefore, unmanageable and inefficient. Clearly, growth has to be accommodated elsewhere within the metropolitan area. High technology sub-centres may emerge in existing or newly created satellite townships, small- to medium-scale manufacturing establishments may locate in the existing industrial suburbs, whereas large scale manufacturing may find new locations on the edge of the metropolis. Here, in a federal structure, the role of the local state becomes important—it identifies or designates industrial or technology parks, and high-tech or export processing zones, and provides basic infrastructure to attract new industry to these locations. Eventually, as the evidence from mega cities around the world seems to indicate, the traditional monocentric city is replaced by larger polycentric urban regions—at an accelerated pace as a result of the reforms (Diniz 1994; Dökmeci and Berköz 1994; Ginsburg, Koppel, and McGee 1991).

THE SPREAD OF INDUSTRY

If unrelenting centralization and concentration of industry was the only reality of economic geography, then convergence between regions would never be possible, and we would not see the rise of new cities after the beginning of industrialization because the initial cities would hold the advantage forever. We know that inter-regional convergence is a reality in several nations, especially in the special cases of the 'new' world: the colonies of Europe in the US, Australia, Canada, etc. Moreover, as we have seen in the preceding sections, policy interventions in favor of lagging regions are implemented almost everywhere. In this section, we provide some examples where industry has relocated—either de-concentrated or re-concentrated—following selected policy interventions.

The unequal division of manufacturing across regions is associated with corresponding wage and land rent differences. Firms considering a move out of the agglomeration would pay lower wages and rents, but forego the benefits of proximity to markets and to intermediate suppliers. Empirical evidence, though scanty, tends to show that spatial decentralization to regions with small employment bases almost never happens. Much more common is de-concentration (what we call 'concentrated decentralization' in Chapter 2) from the core to the periphery of metropolitan areas, and this happens with improvements in the transportation system. For example, Henderson *et al.* (1996) show that many firms moved out of Jakarta to the peripheral areas of the Greater Jakarta metropolitan region in the mid-1980s. These moves were facilitated by the construction of toll ring-roads around the city, retaining some of agglomeration benefits of the region, but reducing congestion costs (for example, land rents and transport costs), enabling firms to benefit from lower land and labour costs in the periphery, which exceeded the increased costs of transportation for serving the same market. Aggregate transport costs per unit of sales revenue also dropped as a larger market could be accessed by a better network. Similarly, for Brazil the de-concentration of industry from Grande Sâo Paulo to lower wage hinterland cities followed the transport corridors first through Sâo Paulo state and then into Minas Gerais, the interior state with the main iron ore and other mineral reserves (Henderson *et al.* 2001).

While the limited evidence shows that inter-regional transport improvements allows firms to relocate from core metro areas to their peripheries, there is no convincing evidence to show relocation to small urban centres in lagging areas. For China, Head and Reis (1996) show that since the implementation of an open door policy in 1978 that ended the prohibition of foreign business and investment, foreign firms preferred to locate in cities with large industrial bases and established foreign investment presence. Their econometric work controls for factors such as provision of fiscal incentives, and availability of infrastructure, which makes some regions inherently more attractive than others. We find a related pattern of re-concentration of manufacturing away from Mexico City to northern cities such as Ciudad Juarez, Monterrey, and Tijuana, which are physically close to the United States following the opening of the Mexican economy to foreign trade and investment (Hanson 1998). Since 1980, industrial activity in Mexico has moved to states on the US–Mexico border, reducing the importance of Mexico City as the nation's main industrial centre. Between 1980 and 1993, the border states increased their share of manufacturing employment from 21 per cent to almost 30 per cent, and Mexico City's share of manufacturing employment declined from over 44 per cent to under 29 per cent. Meanwhile, there have been few new investments in states likes Chiapas, Oaxaca, and Campeche, the lowest income states in Mexico. In Chapter 2, we show that a similar process of concentrated decentralization has taken place in India.

PLAN OF THE BOOK

This is the background against which we locate this examination of industrialization in India. Any serious analysis of industrialization must consider the influence of both the market (that is, economic geography) and the state (that is, political economy). This work integrates the interplay of the market and the state in creating the evolving economic and industrial geography of India. We emphasize the following features:

(1) A spatial perspective: We base our arguments and conclusions on the analysis of detailed data collected at several spatial scales—the state, the metropolis, the district (there are about 500 districts in India), and the postal pin code (there are more than 100 pin codes in each metropolis). This allows us to study the issues identified earlier in this chapter at multiple scales, from the neighbourhood, through the metropolitan region, and the state, to the nation.

(2) The long run: We incorporate data from independence and nationalism into structural reforms and liberalization in the 1990s. For much of the book, we use the structural reforms as the point of departure, but we are always mindful that actions taken today have causes that often go back fifty to 200 years.

(3) Innovative methods and micro-level data: We use current spatial analytical techniques (spatial autoregressive models, Local Indicators of Spatial Association, etc.) in combination with appropriate modelling methods (linear, logistic, and non-parametric models) to tease out the stories from the data. The data we use are disaggregated and detailed—including multi-scale spatial data and firm-level production data—allowing us to build theory from spatial and economic micro-foundations.

(4) Policy orientation: One of our major goals is to inform and improve policy. We analyse how specific policies have had consequences on the ground, and show how the policy framework needs to be changed in order for there to be more balanced, equitable, and sustainable industrial growth.

Chapter 2 'Patterns of Industrial Investment, Old and New' describes and quantifies the spatial distribution of industry at multiple scales (states, districts, metropolitan areas) in the pre-reform and post-reform periods. We also present data from the early post-independence period. We identify and map the emerging economic geography of India and highlight the leading edges and lagging pockets. We show the continuing significance of history (marked by investments in existing industrial areas) and the significance of geography (marked by investments in clusters in and around existing or new industrial areas). We show that the location of post-reform investment favours the coast, advanced regions, and existing metropolises (especially the edge areas); these realities are truer for foreign direct investments than domestic investments (especially the direct investments of the state). The results provide evidence of the return of cumulative causation and divergence.

In Chapter 3 'Determinants of Industrial Location' we identify the factors that influence industrial location decisions. These include variables representing capital, labour, infrastructure, regulation, and geography. We show which of these factors are important and to what extent. We also show that source of capital is the primary cause of spatial divergence in investments. Private capital is profit-oriented and directed towards leading industrial regions, coasts, and metropolises, and away from socialist governments. State industrial investments have some regional equity considerations, and therefore are less biased towards leading regions. The results are established using logistic and OLS (Ordinary Least Squares) regression models on district level data.

In Chapter 4 'Economic Geography and the Firm' we continue the analysis begun in the previous chapter, by looking at what factors influence location decisions at the firm level. More specifically, we ask the question: for manufacturing industry, what are the externalities that matter, and to what extent? We develop an innovative methodology to analyse the influence of economic geography on the cost and wage structure of manufacturing firms. We analyse eight industrial sectors (food/beverages, textiles, leather, printing and publishing, chemicals, metals, machinery, electrical/electronics) by firm size in India, and find that industrial diversity is the only economic geography variable that has a significant, consistent, and substantial cost-reducing effect for firms, particularly small firms. We discuss the implications of these findings for regional growth and development.

In Chapter 5 'Industrial Clusters within Metropolitan Regions', we turn our attention to the internal structure of metropolitan areas and examine the distribution of industry at this scale. The large and growing literature on industrial clustering suggests that firms seek locations that provide localization economies (benefits from having common buyers and suppliers, a specialized/skilled labour pool, and informal knowledge transfers). We show instead that industry location decisions are guided by market imperfections, specifically rigidities in the land market caused by state action (segregationist/environmental policies, the absence of exit policies, and activist industrial promotion policies). We use geographically disaggregated industry location and size data (at the pin code level) from Mumbai, Calcutta, and Chennai, to analyse the eight industry sectors we study in Chapter 4. We test for evidence of global and local clustering, and distinguish between and test for co-clustering and co-location of industries. The results are indicative rather than absolute, and suggest that for location decisions, general urbanization economies are more important than localization economies.

Finally, in Chapter 6 'On Spatial Policy' we summarize the key findings and analyze them in the context of state actions and inactions. We begin by looking at a range of spatial policy initiatives used in international settings. Then we consider the question of why these spatial policies have had such limited effects and identify four categories of answers: (1) that spatial economies are stronger than the fiscal incentives provided by policies; (2) that state policy is fundamentally contradictory—what one hand gives the other takes away; (3) that there is little coordination between policies and that they are ad hoc rather than being based on clear analysis; and (4) that institutions have deep effects that cannot easily be overcome by marginal policies. We end with a discussion on one of the most important questions in contemporary India: can anything be done to bring new investments to Bihar?

NOTE

1. There are many sources that contain basic data on the Indian economy. A reliable one is the portal economywatch.com and its India-related pages. Last accessed 30 August 2006.

2
Patterns of Industrial Investment, Old and New

The objectives of this chapter are two-fold: First, to present the bare bones of the narrative of India's industrialization, beginning in the mid-nineteenth century; second, and more important, to focus on the last decade-and-half by using the structural reforms formally initiated in 1991 as the point of departure in this recent narrative. We fulfill this second objective by presenting aggregate and disaggregated details of the spatial patterns of the pre-reform (or pre-1991) and post-reform (or post-1991) industrial investment, to identify what has changed and the extent of the changes. It is obvious from the discussions in Chapter 1 that we view the structural reforms of 1991 (much more than the hesitant reforms of 1984–5) as a watershed in India's recent political economy with implications not only for industrialization but also for growth, distribution, welfare, culture, and power. Our focus is, of course, on industrialization. Therefore, in this chapter and the next, we use the beginning of the reforms as the critical turning point in recent industrial location and performance.

BACKGROUND

HISTORY OF INDUSTRIALIZATION IN INDIA

The industrial era entered India with the opening of coalmines by Alexander and Company in 1820 and the factory system began with the establishment of the first cotton mill in Bombay by Cowasji Davar in 1851, followed by two more cotton mills by 1860. These mills did not supply the home market, but exported to China and the Far East (Sharma 1954). This was actually a return to the manufacturing

PATTERNS OF INDUSTRIAL INVESTMENT

and trading patterns that existed before the arrival and dominance of the East India Trading Company late in the seventeenth century. Then, India's 'cotton manufactures were in universal demand not only in the Eastern markets from Cairo to Pekin but...in European markets. Indian calicoes were so popular in England in the eighteenth century that...their entry into the English market had to be stopped by legislation' (Sharma 1954: 13; also see Frank 1996).

This is not the place to discuss the details of the economic logic of colonialism, the goals of the imperial project, or the strategies

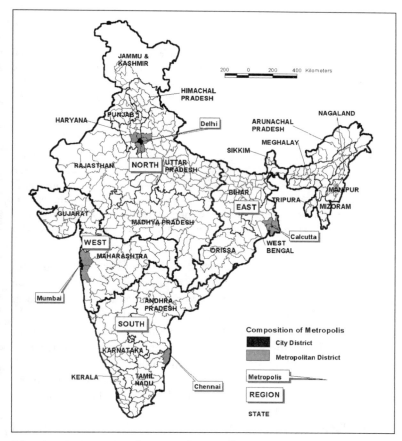

Figure 2.1: Regions, states, districts, and metropolises
Source: Census of India, 1991.

used to implement them. Indian economic history under colonialism is well documented in sources like Bagchi (1976) and Habib (1975). In short, the economic geography created largely by overland trade under various systems of monarchic rule (of empires and kingdoms) was rapidly replaced by an economic geography that was port based. Calcutta, Bombay, and Madras were the three main port cities created and privileged by the company traders, the spearheads used to colonize and then control their respective regions, and these became the centres of industrialization, the economic engines of the suddenly prosperous regions that eventually became the states of West Bengal, Maharashtra, and Tamil Nadu respectively (see Figure 2.1). By the end of the nineteenth century, Digby argued that

> There are two Indias: the India of the Presidency and the Chief Provincial Cities, of the railway system, of the hill stations....There are two countries: Anglostan, the land especially ruled by the English, in which English investments have been made; and Hindustan, practically all of India, fifty miles from each side of the railway lines (Digby 1901: 291–2, quoted in Awasthi 1991).

According to Awasthi (1991: 16–17):

> During 1913–14, the total number of companies in the provinces of Bengal was 973 (35.5%), Bombay 613 (22.3%) and Madras 427 (15.6%). Taken together, these three states...accounted for more than 73 per cent of the companies at work....During 1938–39 Bengal increased its share by six per cent at the expense of Bombay and Madras...when the first Census of Manufacturing was conducted [in 1946] the three major provinces...accounted for 68 per cent of the total factory strength.

Our calculations from data reported in Mitra (1965) show that in early post-independent India, industrialization continued to be regionally unbalanced. The data show that between 1953 and March 1961, of the 4971 industrial licenses granted in the country, about 36 per cent went to the three industrial metropolises—Calcutta, Bombay, and Madras (note these data are just for the metropolitan districts, not the corresponding states or regions). Fully 21.5 per cent of all licenses in the country went to a single district—Bombay and 56 per cent of the licenses were given to firms locating in the three leading states: Maharashtra, West Bengal, and Tamil Nadu.

The slight trend toward de-concentration of industry that could be noticed by the late 1950s became more strongly evident through the following two decades. By the end of the 1970s, and into the early 1980s, industry locations had spread to many states and cities beyond the initial three leaders. Table 2.1 shows the details of formation of fixed capital and employment generation in manufacturing for all regions and selected states and Table 2.2 lists the income and infrastructure indicators for all states with over 1 per cent of India's population. The following points are notable:

(1) Before the institution of reforms in 1991, regional inequality was probably already increasing. In income terms, using any measure of variation, it is possible to show that regional inequality increased between 1960 and 1989. The summary measure, coefficient of

Table 2.1: Long-term trends in industrial formation, 1961–94

	1961	1965	1969	1975	1978	1981	1984	1994
FIXED CAPITAL								
Regions:								
West	29.1	22.9	22.2	25.0	24.3	25.0	25.7	28.1
South	16.4	17.0	22.9	22.2	20.4	19.9	19.4	23.7
North	12.5	23.5	26.4	31.3	30.3	32.1	34.2	28.8
East	37.9	35.1	28.2	19.9	24.6	22.0	20.1	18.5
Selected states:								
Maharashtra	20.4	16.6	15.5	16.3	15.0	16.0	15.9	17.4
Gujarat	8.7	6.3	6.7	8.7	9.3	9.0	9.7	10.7
Tamil Nadu	7.2	8.1	9.9	8.5	7.0	6.8	6.9	8.8
Andhra Pradesh	3.7	3.8	5.8	5.5	5.8	6.1	5.9	9.9
Bihar	12.8	8.0	7.5	6.6	14.1	11.4	9.9	5.1
West Bengal	19.9	19.3	13.7	8.7	6.6	7.1	6.9	8.8
EMPLOYMENT								
Regions:								
West	30.9	28.2	28.0	28.6	26.8	24.8	24.2	23.8
South	22.4	22.2	25.6	26.1	27.5	27.7	27.9	31.4
North	14.9	16.8	19.8	21.7	23.4	24.4	25.9	27.8
East	28.5	30.8	26.1	22.9	21.6	20.6	19.4	15.8
Selected states:								
Maharashtra	20.9	19.6	19.1	19.3	17.7	17.3	16.1	15.1
Gujarat	10.0	8.6	8.9	9.3	9.1	7.4	8.1	8.7
Tamil Nadu	8.2	8.7	10.1	10.0	10.0	10.3	10.4	12.7
Andhra Pradesh	5.7	5.5	6.6	7.3	8.9	8.6	9.3	10.0
Bihar	5.3	5.3	5.1	5.0	5.4	5.0	4.6	3.9
West Bengal	20.2	21.9	17.3	14.5	12.9	12.3	11.4	8.5

Source: Annual Survey of Industries (different years).
Note: All figures are percentages of the total.

Table 2.2: Income and infrastructure indicators, 1950–89

	Per Capita Income India = 100		Installed Power capacity[1]		Road Length in kilometres[1]	
	1950	1989	1961	1989	1961	1989
West						
Gujarat	125.7	130.4	1.7	9.7	13.1	28.3
Maharashtra	123.1	154.0	1.9	10.9	16.6	47.8
North						
Haryana	120.1	160.6	2.7	11.0	28.1	49.5
Madhya Pradesh	77.9	74.7	0.8	5.6	8.5	16.8
Jammu & Kashmir	74.9	80.5	0.6	3.4	4.7	6.9
Punjab	146.5	184.8	1.4	14.5	35.6	65.1
Rajasthan	84.5	84.3	0.3	3.5	12.0	19.5
Uttar Pradesh	70.4	80.5	0.6	3.9	33.6	31.1
South						
Andhra Pradesh	85.2	88.0	0.7	6.2	19.9	43.6
Karnataka	94.7	106.2	0.8	6.8	32.6	52.0
Kerala	100.3	75.3	0.8	5.1	50.2	271.1
Tamil Nadu	80.9	105.6	1.6	5.2	36.0	98.5
East						
Assam	110.6	81.1	0.2	2.1	24.0	66.3
Bihar	59.7	55.7	0.7	2.2	46.6	37.1
Orissa	83.2	75.7	0.7	4.6	20.1	67.8
West Bengal	155.4	100.4	2.1	4.9	73.8	46.3
Coefficient of variation[2]	0.263	0.335	0.608	0.532	0.589	0.966
Population weighted coefficient of variation	0.927	0.999	1.188	1.483	1.170	5.314

Source: Indian Institute of Public Opinion (1993).
Notes: States with populations less than 1% of India's total have not been shown (or used in the calculations). Delhi (in the northern region) is the only important state not listed.
1. The installed power capacity is in Megawatts per 100,000 people; the road length is also per 100,000 people.
2. COV is the Coefficient of Variation; its weighted version is population weighted.

variation, shows increases in both its weighted and un-weighted forms. The income ratio between Punjab (the richest state) and Bihar (the poorest state) rose from under 2.5 to over 3.3. Every high income state, with the exceptions of West Bengal and Assam (both eastern states), raised its income share. However, had we shown regional inequality data for intermediate years, we would have seen that regional inequality declined slightly up to around 1980 and started increasing again. Several scholars (see Awasthi

1991; Chand and Puri 1983; Saha 1987) have documented this. Hence, we can divide the post-independence period into two phases: the first lasted up to 1980 and was characterized by industrial decentralization and declining regional inequality; the second started in the mid-1980s, is still ongoing, and is characterized by increasing regional inequality. We will discuss the industrialization aspects in the remainder of this chapter.

(2) The eastern states had generally declined—in income terms and in the share of industrial investment. Bihar, long the poorest state continued to do poorly, and West Bengal, once the most prosperous state declined to below average levels. In fact, whatever declines in regional inequality took place in earlier decades can be attributed to industrial decline in the east. Further declines, that took them well below the national averages, caused regional inequality levels to increase again. The states in the western region in the meantime raised their income share, and despite some decline in their industrial employment share, improved or continued to hold their positions in most spheres. The states in the northern and southern regions had mixed results, but both regions as a whole made significant gains in industrial capital and employment share.

INDUSTRIAL POLICY AND REFORMS

One of the interesting aspects of early development planning in independent India has been the relative absence of an explicit spatial perspective: 'Indian planning...has been limited to [the] allocation of investment over time, sectors and sub-sectors, whereas there is no explicit spatial dimension in the formal planning models' (Awasthi 1991: 27; also see Chakravarty 1979). The policy makers, led by the Planning Commission and Soviet inspired input-output macroeconomic models, concentrated on the factors of production to the exclusion of any spatial context. Johnson (1970: 167–9) has argued that in addition there were three extenuating factors: one, a belief in the 'redemptive mystique' of large industry, wherever it may be located; two, the 'emulative desire to attain the demographic patterns of the more developed countries'; and, three, what Lewis (1995: 224) calls a 'village fetish.' Lewis argues that two other factors were particularly important: one, that 'any implementation of a positive spatial strategy would cost a lot of money', and two, 'the spatial dimension (of development) was most obviously and blatantly political.' On the other hand, there was poor recognition of the inherently spatial nature of all development; there was a tacit understanding that industrialization would lead to metropolitanization—but that was considered an inseparable part of development, and even desirable.

Some concern over regional disparities was voiced as early as the First Five-Year Plan (1951–6), and emphasized again in the Industrial Policy Resolution of 1956 (which identified the dos and don'ts of Indian industrialization). Over time, a number of instruments to achieve balanced regional industrialization were instituted: the industrial licensing system was used to direct investment into lagging areas, and heavy industry was discouraged (and eventually forbidden) from locating in metropolitan centres; large public sector projects (for example, steel plants,) were located in lagging states like Bihar, Madhya Pradesh, and Orissa; industrial estates, or growth centres, were identified and received some investment in infrastructure; financial incentives for private industrial investment in designated lagging districts (about 60 per cent of all the districts in India) were provided; the prices for 'essential' items such as coal, steel, and cement were equalized nationwide by the Freight Equalization Policy of 1956 (Mills and Becker 1986), which we will discuss in detail in Chapter 6. Despite this, for decade after decade, the industrially advanced states and districts, that is, where the metropolitan centres of Calcutta, Mumbai, and Chennai were located, continued to receive large shares of private investment (Mitra 1965; Kashyap 1979); till 1980 almost 55 per cent of the capital subsidies went to only 25 out of 296 eligible lagging districts, where all 25 were in industrially advanced states. The Freight Equalization policy, meanwhile, 'robbed the producing areas of southern Bihar and Bengal, western Orissa and eastern Madhya Pradesh of their comparative advantage in industries using these products' (Mohan 1983: 51). The National Commission on Urbanization (1988), in delivering the first national level urban policy statement, recommended that most of these polices (with the exception of the growth centre policy) be discontinued.

The structural reforms of 1991 were apparently precipitated by a shortage of foreign exchange reserves, down to about one billion US dollars. Lal (1995) points out that the reaction of 'opening the economy' when faced with a balance of payments crisis had some precedent in India, as in Latin America. It is now clear that the first steps towards liberalization were taken from the time Rajiv Gandhi became prime minister in 1985 (see Kohli 1989). There were important but small steps taken then (as the reforms were gradually overshadowed by political scandal). The reforms announced in 1991 went much further:

> The most striking achievement of the reforms [has been] that commercial considerations, rather than government mandates, are now the determinant in all investment decisions, including ownership, *location*, local content, technology fees, and royalty. The approval authority in the Directorate General of Technical Development in the Ministry of Industries has been eliminated. The Monopolies and Restrictive Practices Act has been amended.... Controls on the import of capital goods have been removed, and the many regulatory bodies (have been) dissolved or reconstructed.... *States now compete with each other to attract new investments* (Dehejia 1993: 88; our emphasis).

Understandably, no new urban or regional policy was announced immediately after the start of liberalization. The restrictions on locating industry in metropolitan areas were lessened. According to the Statement on Industrial Policy of 24 July 1991: 'In respect of cities with population greater than 1 million, industries other than those of a non-polluting nature such as electronics, computer software and printing, will be located outside 25 km of the periphery, except in prior designated industrial areas. A flexible location policy would be adopted in respect of such cities which require industrial regeneration' (quoted in Government of India 1997: 14–15). The Freight Equalization Policy was abandoned, and the central government turned its attention to divesting its interest in the ownership of capital intensive and 'sick' industry. Meanwhile, the Ministry of Urban Affairs and Employment announced that the primary goals of urban development in the new economic order should be cost recovery and replicability of projects (Chakravorty 1996).

In the meantime, most state governments have responded by instituting their own industrial policy reforms. Typically these local reforms have four features: (1) foreign capital and technology is welcomed (special efforts are made to woo non-resident Indian capital); (2) at the state level there is a new 'single window' project clearance agency, which coordinates with district level administrators in matters such as land acquisition; (3) time-bound clearances or sanctions are promised; and (4) environmental hurdles are lowered. In Haryana, for example, project approval and a No Objection Certificate (NOC) from the state Pollution Control Board are promised in fifteen days, and the allotment of land promised in ten days. The state of Punjab tries to grant industrial clearances within 24 hours of application. In Rajasthan, 155 industries were initially exempted from obtaining NOCs from the Pollution Control Board.

At the beginning, though, the reform process was viewed with considerable scepticism by a number of state governments. The socialist government in Bihar and the communist governments in West Bengal and Kerala were especially hesitant, but eventually they too realized that competition between states was going to be fiercer than ever and created new industrial policies to try to take advantage of the new conditions. The chief ministers of all major states have been actively involved in seeking investments, and almost all have led delegations overseas to advertise the investment merits of their respective states.

WHAT CHANGES CAN WE EXPECT?

In order to theorize the effects of the structural reforms, it is necessary to consider two sets of ideas. First, we have to keep in mind the theoretical issues discussed in Chapter 1, principally, the reasons for spatial concentration, outlined in the discussion on economic geography; namely, the influence of market access and increasing returns. Second, we have to consider the proposition that once we strip away all the palliative verbiage justifying the reforms, their main objective is to increase the rate of economic growth. The infamous 'Hindu rate of growth' the nation had managed to eke out in earlier decades had left the national economy lagging far behind export-oriented 'free-traders' like South Korea and Taiwan, nations which had ended the 1940s in economic conditions not dissimilar from India (Ahluwalia 1991). It was felt that the system of licensing, controls, and location incentives and disincentives may have ameliorated regional imbalances to some degree (it is difficult to be sure, specially in light of the contradictions in policy discussed earlier), but they clearly had not boosted growth rates to levels that had been achieved elsewhere. Therefore, the national state (which at that point was an unwieldy coalition government that seemed to be tottering on the verge of collapse) would have to choose a path that appeared to have worked for some nations, a path that was being advocated by the increasingly dominant Washington Consensus discourse. The reformed state must demonstrably disengage from ownership of industry, and from industrial investment location decisions, regulations, and incentives. This is a fundamental condition of structural reform.

We propose three hypotheses on what can be expected for industrial investments in the post-reform stage.

Hypothesis 1: In the short to medium term there will be increased regional divergence because the distribution of new industrial investments can be expected to favour:

(1) advanced industrialized over lagging less industrialized regions;
(2) metropolitan over non-metropolitan regions;
(3) coastal over non-coastal regions.

Hypothesis 2: The location of investments will differ by source of capital—where the private sector investment patterns (especially, the foreign direct investment typically concentrated on high technology, export and consumer goods) will be more strongly oriented toward the coast and metropolitan regions. Conversely, the national state, still fulfilling some obligations to regional equality, will invest more of its increasingly limited resources in inland and non-metropolitan regions than in coastal or metropolitan regions.

Hypothesis 3: The metropolis will undergo an accelerated process of internal restructuring (moving toward polycentricity, or multiple centres) to accommodate the new growth impulses. More capital will be invested in the edge areas than in the over built urban core. This will mitigate the diseconomies of congestion in the large metropolitan areas, and will be helped by state investments in Export Processing Zones (EPZs), technology parks, industrial enclaves, free trade areas, etc. In other words, the state will try to maximize for the private sector the advantages of locating in urban areas by restructuring investments in infrastructure.

In the following pages, we begin to test these hypotheses using data at different spatial scales (region, state, and district), presented in a wide array of tables, graphs, and maps (using new spatial analytical techniques). We begin with a discussion of the data and methods used here. Next, we present the evidence, and follow up with a summary of the findings.

DATA AND METHODOLOGICAL ISSUES

DATA SOURCES

Data on Indian industry is readily available at the state level. However, for the kind of disaggregated analysis required here, state-level data are hardly appropriate. In 1991, the largest state, Uttar Pradesh, had close to 140 million people; the second largest state, Bihar, had over 86 million. If independent, these would be among the largest countries in the world.[1] In fact, it may be reasonable to suggest that almost all Indian analysis should be carried out using some unit smaller than the state. The next largest political unit in the country is the district. There are approximately 500 districts—the number is not fixed (much as the number of states is not fixed) because new districts are carved out of old ones on a fairly regular basis, more so in the 1990s than ever before. Data at the district level, which are generally very difficult to come by, were available for this project. Our analysis is based on two new and unique data sets on Indian industry at the district level.

The Central Statistical Organisation (CSO) has the responsibility of collecting and disseminating most data on the Indian economy. Among the data distributed by the CSO is the Annual Survey of Industries (ASI), which, as the name implies, is an annual publication, at the state level, listing sectorally disaggregated industrial data. These data, which in the past have been made public about eight to ten years after the survey, were never made available at the district level (till we became the insistent and, to our best knowledge, the fortunate first recipients). The ASI covers every factory and manufacturing unit (designated so by the Factories Act of 1948, amended in 1956) using two methods: a 'census' sector survey with 100 per cent coverage of units employing 100 or more persons, and a 'sample' sector survey in which a sample of the smaller units (employing between 20 and 99 persons) is statistically allocated to all districts. The census sector covers over 80 per cent of Indian industry and is considered to be significantly more reliable than the sample sector. We were given district level data from the 1993–4 census sector ASI for six variables (number of factories, fixed capital, invested capital, number of employees, value of output, and net value added). The survey on which these data are based was carried out in 1993, two years after the initiation of reforms. However, since the survey covers every unit that was in operation in 1993, whenever built, this is the most realistic measure of Indian industry for the pre-reform period. Whenever reference is made to pre-reform data, the numbers have been derived from this data set.

There is no government agency tracking new investments after the initiation of reforms. It is widely acknowledged that the best economic data are being generated by a private sector firm: the Centre for Monitoring Indian Economy (CMIE). One of the data sets published quarterly by the CMIE is a list of investment projects with markers for location (state, district, and place), project stage (completed, under implementation, seeking approvals, etc.), product, capital cost, etc. The entire data from January 1992 to February 1998 were collated into a database containing over 10,000 records (the data for 1991, during the middle of which the reforms were announced, were ignored). This database was further whittled down so that it now contains only those projects that have been completed or are under implementation (it is sensible to ignore the projects still in the 'proposal' or 'announcement' phase—no one knows what will happen to them), and those that are not being funded solely by the local government. This final database with about 4650 records forms the basis of all the post-reform calculations.[2]

An explanation of the different geographical scales used here is necessary. In India, regions are typically identified with states; in other words, studies of regional development or inequality in India have always used the state as the unit of analysis; for example, Maharashtra and Punjab are considered advanced regions while Bihar and Orissa are considered lagging regions. With the district level data used here, it is possible to create new data driven definitions of advanced and lagging regions that are distinct from politically defined regional boundaries. However, such definitions may be questionable (as somewhat arbitrary parameters will have to be used), and not very meaningful for India scholars who are used to dealing with information at the state level, nor very meaningful for policy makers whose actions are designed for the state level (whether or not the policies are being made at the centre). Moreover, the only possible method of presenting a historical account of comparative regional change in India, is to begin from the unit of the state. Nevertheless, as pointed out earlier, there are significant problems inherent in using only large and often incompatible political units for analysis.

Hence, the data are presented at four spatial scales (all the scales used in this chapter are shown in the political map of India in Figure 2.1):

(1) One type of meta region (the geographical groupings of states: east, west, north, south, already used earlier in this chapter);
(2) Two types of meso regions:
   (i) The political region (identical to the state), and
   (ii) The metropolitan region (or agglomerations of districts that form commonly recognized metropolitan units);
(3) One type of micro region (the district). In Chapter 5, we will introduce an even smaller enumeration unit—the pin code—but that is not necessary at this stage.

Finally, a note on the data aggregation scheme used here is necessary at this point. The ASI data were reported at the 2-digit National Industrial Classification (or NIC) code. The CMIE data are project data and do not have NIC codes. The two data sets were made compatible using the following method: First, the ASI data, which are reported in 26 NIC 2-digit codes, were further aggregated into six sectors: heavy industries, chemicals and petroleum, textiles, agribusiness, utilities, and other (as shown in Table 2.3). The analysis was carried out for all industry (that is, all 26 NIC 2-digit categories combined together), and for the first five of the sectors identified here (the 'other' category, which is not internally coherent, was not used for analysis). The ASI format was then used to assign general classification codes to the CMIE listed (or post-reform) projects. In the following sections the results for all industry and the five sectors are separately presented wherever possible.[3]

Table 2.3: Industrial classification codes and aggregation methods

Category used in this study	NIC Code	Description
Agribusiness	20-21	Manufacture of Food products
	22	Manufacture of Beverages, Tobacco and related products
Textiles	23	Manufacture of Cotton Textiles
	24	Manufacture of Wool, Silk and Man-Made Fibre Textiles
	25	Manufacture of Jute and other Vegetable Fibre Textiles (except cotton)
	26	Manufacture of Textile Products (including wearing apparel)
Chemicals and petroleum	30	Manufacture of basic Chemicals and Chemical Products (except products of petroleum or coal)
	31	Manufacture of Rubber, Plastic, Petroleum and Coal Products, and Processing of Nuclear Fuels
Heavy industry	32	Manufacture of Non-Metallic Mineral Products
	33	Basic Metal and Alloys Industries
	34	Manufacture of Metal Products and Parts, except machinery and equipment
	35-36	Manufacture of Machinery and Equipment other than transport equipment
	37	Manufacture of Transport Equipment and Parts
	38	Other Manufacturing Industries
Utilities	40	Electricity
	41	Gas and Steam Generation and Distribution through Pipes
	42	Water Works and Supply
	43	Non-Conventional Energy Generation and Distribution
Other (Not analysed here)	27	Manufacture of Wood and Wood Products
	28	Manufacture of Paper and Paper Products and Printing, Publishing and Allied Industries
	29	Manufacture of Leather and Products of Leather, Fur and substitutes of Leather
	39	Repair of Capital Goods
	74	Storage and Warehousing Services
	97	Repair Services

MAPPING METHODS

We have used two types of maps. Both are choropleth maps. The first displays a conventional variable—the quantity of investment—standardized by the average national investment. The second type of map uses an analytical device called the G_i^* statistic (from Getis and Ord 1992). Before describing the properties of this statistic, it may be useful to discuss briefly why it is necessary. The first law of geography states that everything is related to everything else, but events or phenomena closer together are more closely related. This feature of spatial distributions is called spatial autocorrelation which is positive when like values cluster together (that is, high values are proximate to high values, and low values are proximate to low values) which is a clear expectation of geographers, or negative (when high and low values are proximate) which is rare and difficult to explain. There are several tests of spatial autocorrelation, the most well known of which is Moran's I. Formally, Moran's I is given by:

$$I = \frac{N}{S_0} \frac{\sum_i \sum_j w_{ij} \cdot (x_i - \mu) \cdot (x_j - \mu)}{\sum_i (x_i - \mu)^2}$$

where,
N is the number of cases or observations,
w_{ij} is the element in the spatial weights matrix corresponding to the observation pair i, j,
x_i and x_j are observations for locations i and j (with mean μ),
and S_0 is a scaling constant (that is, the sum of all spatial weights).

The weights matrix can be constructed in several ways: the contiguity weights matrix assigns a weight of 1 to all j that are contiguous to i (that are direct neighbours), and a weight of 0 to all other j (note that it is possible to construct higher order contiguity weights matrices). A distance weights matrix assigns a value of 1 to all j that are within a chosen critical distance of x, and a value of 0 to all other j. For instance, if 150 kilometres is chosen as the critical distance (as it is here), all districts whose centroids lie within 150 kilometres of the centroid of district x are assigned a spatial weight of 1. The theoretical mean of Moran's I is $-1/(N-1)$, which is nominally negative but tends to 0 when N increases. The data were tested for spatial autocorrelation using Moran's I with a first order spatial weights matrix, and contiguity distances of 100 kilometres and 150 kilometres (discussed in detail later).

However, Moran's I is a test of global spatial auto-correlation, useful in indicating whether a whole system is clustered. It is less useful in identifying areas of local spatial association or dependence where there may be no clustering effects compared to a global mean, but they may very well show spatial dependence when compared to a local mean (see Anselin 1995). According to Anselin (1992: 23-2) 'these statistics allow for the decomposition of a global measure of spatial association into its contributing factors, by location. They are thus particularly suitable to detect potential non-stationarities in a spatial data set, for example, when the spatial clustering is concentrated in one subregion of the data only." We used the G_i^* statistic which is formally given by:

$$G_i^* = \frac{\sum_j w_{ij}(d) x_j}{\sum_j x_j}$$

Where, $w_{ij}(d)$ are the elements of the contiguity matrix for distance d. Inference about the G_i^* statistic is based on the standardized z-value (which is computed by subtracting the theoretical mean and dividing by the theoretical standard deviation). Positive and significant z-values imply clusters of high values, whereas negative and significant values imply clusters of low values (unlike the interpretation of z-values for Moran's I).

FINDINGS

The empirical findings are reported in three sections. Note that though the theoretical expectations were presented in the form of a set of hypotheses, the focus here is not on formal hypothesis testing. The data being presented here are quite detailed and complex—the data are more detailed than any that have been used before for studying Indian regional development, perhaps for the regional study of any developing nation; they are also complex because a major element of the study is spatial analysis, which requires methods that are necessarily different from conventional econometrics. Hence, a mix of methods has been used to try to get to the core findings. The first section is a presentation of tabular and graphical data usually in aggregated form for the meta, meso and micro regions; in the third section, we use maps to present micro region (or district) level data.

DISTRIBUTION BETWEEN STATES AND DISTRICTS

Table 2.4 presents data on an important manifestation of reforms—that is, whether there is indeed any decline in the nation-state's share of industrial ownership. The numbers make the case quite clearly: the share of the public sector (the central, state, and local governments) reached a peak in about 1985–6, or about the time pro-market Rajiv Gandhi became prime minister, in every category—fixed capital, employment, and net value added. Correspondingly, the share of the private sector was lowest in the early 1980s and has risen significantly since that time, with the most dramatic increase being seen in the share of fixed capital. One cannot say anything very interesting about the joint sector (that is public–private partnerships) except that its heyday came, like for the public sector, in the early 1980s. Also noteworthy is the fact that the public sector is much more capital intensive than the private sector; the latter employs about 2.5 times more workers per unit of fixed capital.

HYPOTHESIS 1: LEADING REGIONS VERSUS LAGGING REGIONS

Consider the hypothesis that the already advanced regions and states (that is meta regions and political regions) would be the prime beneficiaries of new investment (hypothesis 1a). Figures 2.2 and 2.3 show the pre- and post-reform distribution of investment on a per capita basis for Indian regions and states. Table 2.5 contains summary and sectorally disaggregated data (following the categories listed in Table 2.1) on pre- and post-reform investment at the state level.[4] These data are somewhat inconclusive, even surprising on occasion.

Clearly the western region, already the leading meta region before the initiation of reforms, has gained the most, and the two western states, Gujarat and Maharashtra, are individually the two top investment destinations (by quantity). However, Maharashtra, which had been ranked first in total investments and second in per capita investments has lost its rankings and its share. It is now the second state in total investment terms, and fourth in per capita terms. Equally clearly, the two lagging meta regions—the rapidly declining east, and what had been the improving north in the pre-reform phase, are also the least able to attract post-reform investments. West Bengal and Bihar, the two largest states in the east, are also two of the least

Table 2.4: Industrial investment by type of ownership, 1973–95

	Fixed Capital				Employment				Net Value Added			
	1973–74	1985–86	1989–90	1994–95	1973–74	1985–86	1989–90	1994–95	1973–74	1985–86	1989–90	1994–95
Public	60.1	61.6	55.0	43.3	23.5	29.6	27.4	24.2	23.0	33.2	30.1	25.9
Joint	5.6	10.2	7.5	9.5	3.5	7.0	6.6	4.7	3.9	9.8	9.1	6.6
Private	34.3	28.2	37.5	47.2	73.0	63.4	66.0	71.1	73.1	57.0	60.8	67.5

Source: Annual Survey of Industries (different years).

Table 2.5: Statewise distribution of investment in selected sectors, pre- and post-reform

STATE	All Industry		Heavy Industry		Chemicals		Utilities	
	Pre-Reform Share of National Total (%)	Post-Reform Share of National Total (%)	Pre-Reform Share of National Total (%)	Post-Reform Share of National Total (%)	Pre-Reform Share of National Total (%)	Post-Reform Share of National Total (%)	Pre-Reform Share of National Total (%)	Post-Reform Share of National Total (%)
Andhra Pradesh (AP)	8.85	6.22	11.30	5.29	6.04	5.81	9.32	8.63
Assam (ASS)	0.75	2.19	0.14	0.12	1.28	4.98	0.03	1.47
Bihar (BIH)	6.09	3.74	12.24	7.33	2.04	0.90	4.81	3.04
Chandigarh (CHN)	0.03	0.05	0.03	0.10	0.00	0.00	0.00	0.00
Dadra & Nagar Haveli	0.09	0.25	0.26	0.31	0.03	0.09	0.00	0.00
Delhi (DLH)	1.00	0.15	0.70	0.03	0.08	0.03	1.46	0.07
Goa	0.29	0.27	0.22	0.44	0.61	0.13	0.07	0.16
Gujarat (GUJ)	10.05	16.89	5.60	7.80	24.43	30.07	6.84	14.98
Haryana (HAR)	2.53	2.36	4.25	2.63	0.48	3.11	1.86	1.03
Himachal Pradesh (HP)	0.79	1.33	0.52	0.71	0.16	0.06	1.33	3.10
Karnataka (KAR)	3.66	7.17	4.96	8.69	1.29	2.76	3.02	10.02
Kerala (KER)	1.96	1.30	0.72	0.68	3.57	1.53	2.46	2.13
Madhya Pradesh (MP)	6.33	7.94	8.86	11.66	3.37	5.78	5.85	4.99
Maharashtra (MAH)	18.11	13.44	16.42	14.98	24.05	13.89	14.59	10.07
Orissa (ORI)	4.67	5.96	7.12	11.85	2.08	1.19	5.45	4.60
Pondicherry (PON)	0.21	0.37	0.27	0.20	0.22	1.04	0.00	0.02
Punjab (PUN)	3.86	3.94	2.16	1.57	2.25	6.73	5.50	0.32
Rajasthan (RAJ)	3.89	3.71	3.30	4.02	4.05	0.97	3.88	5.75
Tamil Nadu (TN)	8.48	8.71	5.27	9.46	9.51	8.38	9.03	9.61
Tripura (TRI)	0.03	0.10	0.01	0.04	0.00	0.00	0.08	0.33
Uttar Pradesh (UP)	10.68	8.15	5.39	6.32	10.07	8.13	15.67	11.04
West Bengal (WB)	7.30	5.64	10.16	5.59	4.05	4.43	8.17	8.24

Source: Authors' calculations from the ASI and CMIE data. Discussed in text.

Table 2.6: Top 25 districts and their shares, pre- and post-reform

	All Industry				Heavy Industry			
	Pre-Reform		Post-Reform		Pre-Reform		Post-Reform	
1	Greater Bombay MAH	8.23	Bharuch GUJ	4.29	Vishakhapatnam AP	7.61	Barddhaman WB	3.95
2	Vadodara GUJ	3.90	Surat GUJ	4.28	Barddhaman WB	7.50	Raigarh MAH	3.93
3	Lucknow UP	3.71	Jamnagar GUJ	3.55	Purbi Singhbhum BIH	6.73	Ganjam ORI	3.55
4	Vishakhapatnam AP	2.93	Chengaianna TN	3.45	Pune MAH	5.02	Chengaianna TN	3.45
5	Barddhaman WB	2.74	Raigarh MAH	2.82	Sundargarh ORI	4.42	Pune MAH	3.43
6	Madras TN	2.48	Greater Bombay MAH	2.75	Greater Bombay MAH	4.26	Dakshin Kannad KAR	2.89
7	Pune MAH	2.34	Dakshin Kannad KAR	2.57	Dhanbad BIH	4.08	Raipur MP	2.67
8	Hyderabad AP	2.31	Vishakhapatnam AP	2.06	Durg MP	2.98	Koraput ORI	2.63
9	Purbi Singhbhum BIH	2.26	South Arcot TN	2.00	Bangalore KAR	2.56	Bellary KR	2.62
10	Patiala PUN	1.81	Ratnagiri MAH	1.80	Gurgaon HAR	1.83	Ghaziabad UP	2.56
	Top 10 Total:	32.71	Top 10 Total:	29.58	Top 10 Total:	46.99	Top 10 Total:	31.68
11	Raigarh MAH	1.80	Ghaziabad UP	1.75	Chengaianna TN	1.52	Surat GUJ	2.50
12	Bangalore KAR	1.79	Bhatinda PUN	1.68	Raigarh MAH	1.27	Purbi Singhbhum BIH	2.38
13	Jabalpur MP	1.67	Pune MAH	1.61	Faridabad HAR	1.24	Sundargarh ORI	2.14
14	Surat GUJ	1.66	Thane MAH	1.61	Surat GUJ	1.21	Dhanbad BIH	1.86
15	Chengaianna TN	1.52	Barddhaman WB	1.52	Aurangabad MAH	1.21	Gurgaon HAR	1.84
16	Sundargarh ORI	1.50	Bellary KAR	1.39	Bharuch GUJ	1.20	Bharuch GUJ	1.76
17	N. 24 Parganas WB	1.50	Ganjam ORI	1.36	Ghaziabad UP	1.19	South Arcot TN	1.67
18	Dhanbad BIH	1.44	Medinipur WB	1.35	Thane MAH	1.14	Durg MP	1.57
19	Sonbhadra UP	1.38	Vadodara GUJ	1.34	Salem AP	1.03	Ratnagiri MAH	1.47
20	Dhenkanal ORI	1.36	Sagar MP	1.15	Sonbhadra UP	0.95	Kedunjhar ORI	1.36
21	Thane MAH	1.33	Cuttack ORI	1.11	Bilaspur MP	0.91	Gulbarga KAR	1.24
22	Bharuch GUJ	1.20	Dibrugarh ASS	1.04	Medak AP	0.90	Vishakhapatnam AP	1.15
23	Jaipur RAJ	1.17	Purbi Singhbhum BIH	1.04	Raipur MP	0.85	Kachchh GUJ	1.14
24	Durg MP	1.02	Raipur MP	1.03	Chittaurgarh RAJ	0.81	Thane MAH	1.11
25	Ahmadabad GUJ	0.90	Koraput ORI	0.98	Nagpur MAH	0.80	Sidhi	1.08
	Top 25 Total:	53.95	Top 25 Total:	49.53	Top 25 Total:	63.22	Top 25 Total:	55.94

Table 2.6 (contd)

	Chemicals & Petroleum			Utilities		
	Pre-Reform		Post-Reform		Pre-Reform		Post-Reform	
1	Vadodara GUJ	12.41	Jamnagar GUJ	12.36	Greater Bombay MAH	14.47	Surat GUJ	7.16
2	Greater Bombay MAH	8.17	Bharuch GUJ	7.81	Lucknow UP	11.98	Bharuch GUJ	4.25
3	Raigarh MAH	7.14	Greater Bombay MAH	7.01	Madras TN	7.48	Thane MAH	4.21
4	Chengaianna TN	4.68	Bhatinda PUN	6.23	Hyderabad AP	7.27	Dakshin Kannad KAR	3.63
5	Thane MAH	3.94	Chengaianna TN	5.43	Patiala PUN	5.49	Garhwal UP	3.28
6	Bharuch GUJ	3.85	Sagar MP	4.26	Jabalpur MP	5.35	South Arcot TN	2.97
7	Ernakulam KER	2.88	Surat GUJ	4.21	Vadodara GUJ	4.74	Vishakhapatnam AP	2.87
8	Kota RAJ	2.61	Medinipur WB	4.11	N. 24 Parganas WB	3.49	East Godavari AP	2.41
9	Budaun UP	2.47	Vishakhapatnam AP	3.44	Dhenkanal ORI	3.34	Ghaziabad UP	2.18
10	East Godavari AP	2.19	Vadodara GUJ	2.89	Jaipur RAJ	3.31	Chengaianna TN	2.16
	Top 10 Total	50.33	Top 10 Total	57.75	Top 10 Total	66.93	Top 10 Total	35.10
11	Pune MAH	2.18	Ratnagiri MAH	2.89	Sonbhadra UP	3.27	Murshidabad WB	2.08
12	Surat GUJ	2.05	Raigarh MAH	2.83	Murshidabad WB	2.84	Sidhi MP	2.01
13	Vishakhapatnam AP	1.92	Dibrugarh ASS	2.65	Thiruvananthapuram TN	2.43	Ratnagiri MAH	1.90
14	Chidambaranar TN	1.86	Panipat HAR	2.40	Patna BIH	2.19	Cuttack ORI	1.85
15	Cuttack ORI	1.40	South Arcot TN	2.29	Bangalore KAR	2.01	Greater Bombay MAH	1.62
16	Jamnagar GUJ	1.35	Dakshin Kannad KAR	2.14	Karimnagar AP	2.00	Tirunelveli TN	1.57
17	Junagadh GUJ	1.25	Mathura UP	1.52	Gaya BIH	1.88	Rai Bareli UP	1.53
18	Mahesana GUJ	1.21	Ernakulam KER	1.50	Ambala HAR	1.82	Uttar Kannad KAR	1.51
19	Medinipur WB	1.15	Etawah UP	1.47	South Arcot TN	1.53	Madras TN	1.47
20	Mathura UP	1.13	Sibsagar ASS	1.30	Surat GUJ	1.52	Raigarh MAH	1.43
21	Valsad GUJ	1.12	Cuttack ORI	1.05	Medinipur WB	1.30	Dholpur RAJ	1.43
22	Hugli WB	1.02	Nellore AP	1.01	Puri ORI	1.24	S. 24 Parganas WB	1.36
23	Nilgiri TN	1.00	Shahjahanpur UP	0.84	Shimla HP	1.18	Dibrugarh ASS	1.36
24	Moradabad UP	0.91	Kachchh GUJ	0.77	Ahmadabad GUJ	0.58	Chamba HP	1.35
25	Bhind MP	0.88	Bareilly UP	0.76	Raichur KAR	0.58	Hugli WB	1.34
	Top 25 Total:	70.77	Top 25 Total:	83.20	Top 25 Total:	93.31	Top 25 Total:	58.91

Source: Authors' calculations from the ASI and CMIE data. Discussed in text.

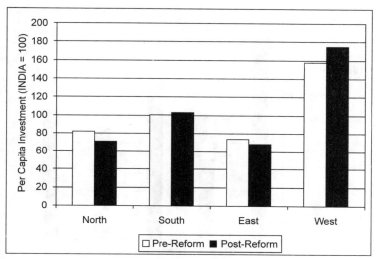

Figure 2.2: Investment distribution by region

Source: Pre-reform data from ASI (data files for 1993–4); post-reform data compiled from CMIE (different years); authors' calculations. Details discussed in text.

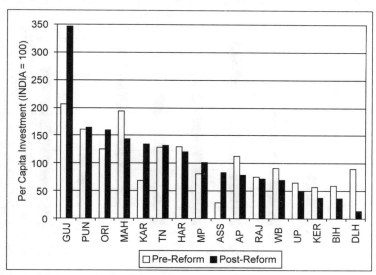

Figure 2.3: Investment distribution by state

Source: Pre-reform data from ASI (data files for 1993–4); post-reform data compiled from CMIE (different years); authors' calculations. Details discussed in text.

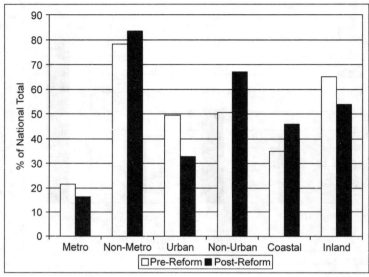

Figure 2.4: Investment distribution by metropolitan, urban, and coastal attributes

Source: Pre-reform data from ASI (data files for 1993–4); post-reform data compiled from CMIE (different years); authors' calculations. Details discussed in text.

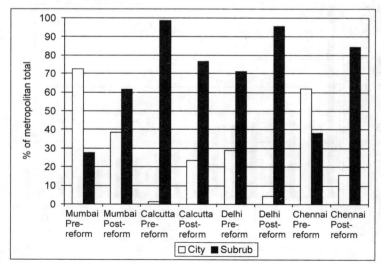

Figure 2.5: Investment distribution by city and suburb

Source: Pre-reform data from ASI (data files for 1993–4); post-reform data compiled from CMIE (different years); authors' calculations. Details discussed in text.

Sectoral Composition

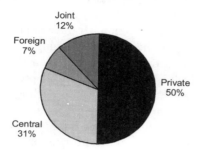

Coastal Investment

Inland Investment

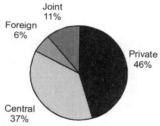

Metropolitan Investment

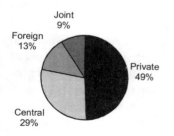

Non-Metropolitan Investment

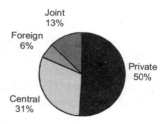

Figure 2.6: Investment distribution by capital source

Source: Calculated from CMIE (different years); authors' calculations. Details discussed in text.

sought after investment destinations. But Orissa, another eastern state that had earlier been largely bypassed by private capital (the majority of the pre-reform investments there were by the central government) has unexpectedly moved up to number three in per capita post-reform investment. Similarly, in the north, the lagging states of Uttar Pradesh and Rajasthan have attracted even less than their already meagre share of capital; Delhi's share has fallen off precipitously, and even prosperous Haryana has seen its share decline. Only Punjab has managed to hold on to and even improve its share position marginally. The position of the south as a whole has barely changed, but within the region, the data show significant gains in Karnataka, marginal gains in Tamil Nadu, and significant losses in Andhra Pradesh (earlier an above average state), and Kerala (which had already been doing very poorly).

At the state level, it is easy to identify the states that have improved their positions significantly: Gujarat, Orissa, Karnataka, and Assam (from all four meta regions), and their opposite numbers, the states whose positions have worsened significantly: Maharashtra, Haryana, Delhi, Uttar Pradesh, Andhra Pradesh, Kerala, and West Bengal (again from all four meta regions). Among these, Maharashtra is clearly the most affected (though it is necessary to point out again that it is still a leading investment magnet) with losses in all sectors.

Now consider hypothesis 1b (on the performance of metropolitan regions) using the district-level data in Table 2.6, beginning with the category of All Investment. Only two of the top ten districts from the pre-reform period have managed to remain in the top ten in the post-reform period. These are Greater Bombay and Vishakhapatnam, both of which have lost share in the transition. Greater Bombay's experience is illuminating: it was by far the leading district in the country in the pre-reform period (with 8.23 per cent of the national fixed capital); after the reforms it is ranked 6th, with 2.75 per cent of the national total. Greater Bombay's loss more than accounts for Maharashtra's total loss; in fact, ignoring Greater Bombay, the rest of Maharashtra has actually seen an increase in its share of investment. This is an important development, and its implications will be discussed elsewhere.

At the same time, some pre-reform top ten districts have dropped out of the post-reform top 25 list altogether: these are Madras,

Hyderabad, Lucknow, and Patiala, all urban districts with the first three being the core of the fourth, fifth, and tenth largest metropolises in India. The ranking of districts in the heavy industry sector alone (comprising about one-third of the total investment in both pre-and post-reform phases) shows a similar turnover. It appears that the metropolitan districts have declined.

This finding for individual metropolitan districts is confirmed for all metropolitan districts considered together in Figure 2.4.[5] Their share of investment has declined from almost 23 per cent in the pre-reform period to under 18 per cent in the post-reform period. Similarly, the share of urban districts as a whole (where urban districts are identified as those that are over 50 per cent urban) has declined substantially—from just under 50 per cent to about 32 per cent.

Simultaneously one can note the rise of non-metropolitan areas (in Table 2.6). Greater Bombay, Chennai, Delhi, Hyderabad, Ahmedabad, Bangalore, Lucknow, etc., have all declined; Calcutta has improved marginally, but from an extremely poor position and still cannot come close to breaking into the top 25 list. In contrast, some suburban districts have risen—such as Chengaianna (surrounding Chennai), and Raigarh and Thane (around Mumbai). But the most impressive performance has been shown by non-metropolitan, even non-urban districts such as Bharuch, Jamnagar, Dakshin Kannad, and South Arcot. This list appears to show the emergence of India's new economic geography—a leading edge of non-metropolitan, coastal districts that are in close proximity to metropolitan areas.

Turning now to hypothesis 1c, the data show the continuing decline of the inland region. Every district in the top ten after the reforms lies to the south of the Vindhyas, the somewhat imaginary line dividing north and south India. Uttar Pradesh and Bihar together account for a quarter of India's population, yet have only one district each in the post-reform top 25: Ghaziabad in Uttar Pradesh, which is part of the Delhi metropolis, and Purbi Singhbhum, the manufacturing base of the giant Tata group. The heavy industry list has more inland districts, most notably Bardhamaan (the leader), where a large proportion of the new investment is by the central government in existing iron and steel plants. Figure 2.4 confirms the theoretical expectation regarding coastal versus inland districts. The share of the former has increased by 10 percentage points, going from under 35 per cent to over 45 per cent of total investment, an increase of unexpected magnitude.

HYPOTHESIS 2: DOES CAPITAL SOURCE MAKE A DIFFERENCE?

A sectoral disaggregation of the post-reform investment (Figure 2.6) is informative. First note some of the more interesting aspects of the composition of total new investment: (1) the share of the central government appears to be continuing the downward trend shown in Table 2.4; that is, the declining involvement of the nation-state in the ownership of industry continues.[6] (2) The share of FDI (7 per cent) is a rather small fraction of the total new investment. (3) The share of the joint sector (12 per cent) is higher than at any earlier time. (4) The share of the private sector has increased to 50 per cent, also higher than at any earlier time (in fact, adding the FDI share, which is also private capital, raises the total private share even higher).

We hypothesized that foreign capital would tend to be more concentrated in coastal and metropolitan districts. The data in Figure 2.6 lend support to this hypothesis: We see that the share of FDI in all coastal investment is 50 per cent more than in all inland investment; similarly, it is more than double in metropolitan versus non-metropolitan districts. The share of private investment in coastal areas (not including FDI) is about 2.5 times that of the central government. For coastal areas the share of government investment is in the opposite direction from that of FDI and private investments; 37 per cent of inland versus 23 per cent of coastal investments. The sectoral decomposition shows little difference for government and private investment between metropolitan and non-metropolitan areas. We take up the issue of the location impact of capital source in greater detail and with more analytical rigour in the next chapter.

HYPOTHESIS 3: DO INVESTMENTS FAVOUR THE METROPOLITAN EDGES OVER THE CORES?

The data in Figure 2.5 show general conformity with expectations (and mirror the information in Table 2.6, where we have already seen evidence of the decline of core metropolitan districts, and the rise of some districts bordering the core districts). We suggested that the new metropolitan investments would be located outside the original city area, in the (for lack of a better term) suburban or peri-urban districts. In Figure 2.5 we see that in three of the four cases listed, the suburban districts received more investment than the city districts. The only exception is Calcutta, which has gained relative to its suburban districts in the post-reform period. In the Mumbai metropolitan area, the suburban districts have gained more in the post-reform period, and spectacularly more so in suburban Chennai and Delhi.

SPATIAL DISTRIBUTIONS AND CLUSTERS

The preceding section has provided some clear indications on the spatial changes in India's industrial economy as a result of (or after) the liberalizing reforms. However, the tabular data cannot show the relative locations of the lagging/advanced districts, neither can tables provide information on the extent of spatial clustering. This section continues the analysis using two spatial devices discussed in the previous section—a measure of the extent of clustering (using Moran's I), and maps of significant clusters of investment and lack of investment using the G_i^* statistic. Moran's I measures whether spatial autocorrelation exists, and G_i^* is an indicator of local spatial association which can also be disaggregated by location and mapped (Getis and Ord 1992).

THE EXTENT OF SPATIAL CLUSTERING

The top district lists seen in Table 2.6, while informative, do not include the full distribution of investments; neither is it possible to make summary judgments on the extent of concentration or clustering or change in either dimension. It is important at this point to distinguish between the concepts of 'concentration' and 'clustering'. There is an emerging literature on industry concentration and its relationship to agglomeration, trade, and growth. This area has long been of interest to urban geographers and urban economists, and over the past decade has received renewed interest following the work of Krugman (1991; also see Ellison & Glaeser 1997; Rosenthal & Strange 2001). Several devices to measure industry concentration have been discovered or rediscovered. Prominent among these are the 'spatial Gini' and Ellison and Glaeser's γ. These measures suffer from a common drawback, one that White (1983) termed the 'checkerboard problem', whereby these measures are not really spatial—any geographical arrangement of parcels (in this case districts) would yield the same measure of concentration. Hence 'concentration' has to be distinguished from 'clustering' where the latter is explicitly spatial; that is, geographical arrangements are incorporated in measures of clustering, which is a case of spatial auto-correlation.

The summary measures reported in Table 2.7 use the Gini coefficient to measure concentration (γ was also calculated, but since the number of observations is high, it is virtually indistinguishable

Table 2.7: Measures of industrial concentration and clustering, pre- and post-reform

	Gini Coefficient		Moran's I—Contiguity		Moran's I—Distance	
	Pre-Reform	Post-Reform	Pre-Reform	Post-Reform	Pre-Reform	Post-Reform
All Industry	0.732	0.706	0.093	0.161	0.173	0.183
			(3.28)	(5.65)	(7.29)	(7.72)
Heavy Industry	0.808	0.774	0.090	0.160	0.127	0.128
			(3.17)	(5.61)	(5.36)	(5.42)
Chemicals and Petroleum	0.864	0.930	0.153	0.063	0.234	0.068
			(5.35)	(2.27)	(9.83)	(2.93)
Textiles	0.817	0.899	0.104	0.147	0.096	0.170
			(3.65)	(5.14)	(4.09)	(7.18)
Agribusiness	0.662	0.711	0.352	−0.001	0.304	0.002
			(12.23)	(0.07)	(12.74)	(0.20)
Utilities	0.958	0.848	−.0008	0.104	0.054	0.089
			(0.18)	(3.67)	(2.36)	(3.81)

Source: Pre-reform data from ASI (data files for 1993–4); Post-reform data compiled from MIE (different years; authors' calculations. Details discussed in text.
Notes: The Gini coefficient is a measure of concentration and is calculated from

$$G_i = 1 + 1/m - 2/m \Sigma j_{ik} * (\Sigma r_{ik} p_{ik})$$

Where, G_i = Gini coefficient for industry i, m = number of districts, r_{ik} = the rank of district k for industry i, p_{ik} = the investment share of district k for industry i.
Moran's I is a measure of clustering. The derivation of I is in the main text. The figures in parenthesis are the Z-scores of the I-values.

PATTERNS OF INDUSTRIAL INVESTMENT

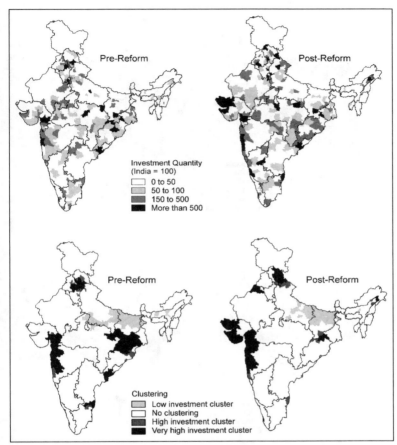

Figure 2.7: Distribution and clustering of total investment on a per-capita basis

Source: Pre-reform data from ASI (data files for 1993–4); post-reform data complied from CIME (different years); authors' calculations. Details discussed in text.
Note: Maps not to scale.

from the Gini), and Moran's I to measure clustering. The results are very interesting. In general, concentration and clustering processes have moved in opposite directions; that is, where concentration has decreased clustering has increased, and vice versa. Concentration has declined for all industry, heavy industry, and utilities, and it has increased for chemicals and petroleum, textiles, and agribusiness. Clustering has increased for all but chemicals and agribusiness; for

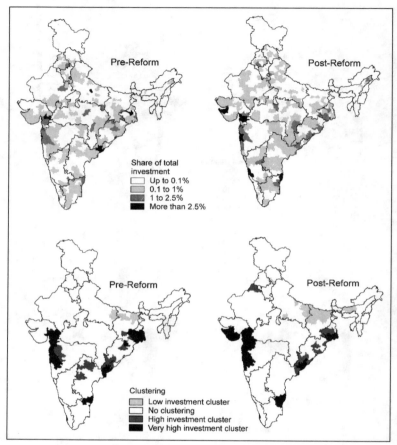

Figure 2.8: Distribution and clustering of investment as share of national total

Source: Pre-reform data from ASI (data files for 1993–4); post-reform data compiled from CMIE (different years); authors' calculations. Details discussed in text.
Note: Maps not to scale.

the latter, there is no evidence of clustering in the post-reform data.[7] The textiles sector is the only one where both measures have moved in the same, higher direction. Consider the implications of the general finding by looking at one instance, say heavy industry. The decline in concentration is to be expected from the data in Table 2.6. The top districts have lost share, therefore lower ranked districts have higher

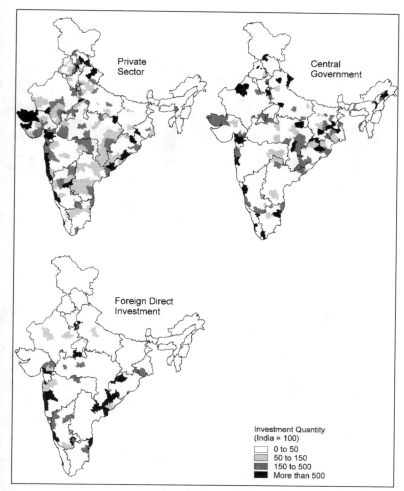

Figure 2.9: Distribution of investment by capital source

Source: Calculated from CMIE (different years); authors' calculations. Details discussed in text.
Note: Maps not to scale.

shares than before. The increase in clustering suggests that the spread of heavy industry investment is greatest in districts which are adjacent or proximate to the existing high investment districts. In other words, the new investments are spatially more concentrated than before.

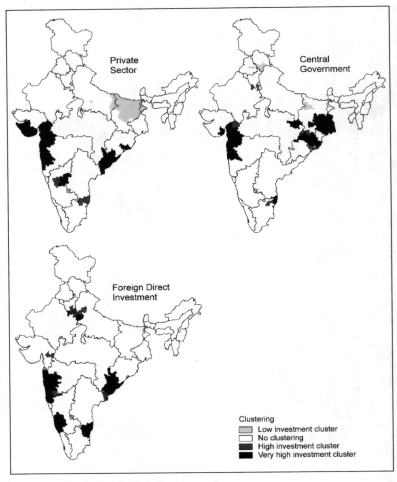

Figure 2.10: Clustering of investment by capital source

Source: Calculated from CMIE (different years); authors' calculations. Details discussed in text.
Note: Maps not to scale.

WHAT THE MAPS SHOW

Each of the figures in this section, with the exception of Figures 2.9 and 2.10, has four maps. The pair on top are choropleth maps showing the distribution of investment in a standardized form, where the nationwide investment is standardized at 100 (the only minor

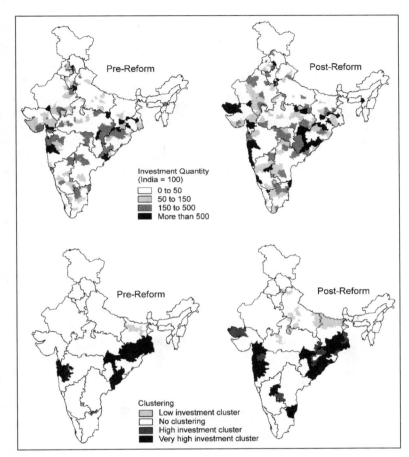

Figure 2.11: Distribution and clustering of heavy industry

Source: Pre-reform data from ASI (data files for 1993–4); post-reform data compiled from CMIE (different years); authors' calculations. Details discussed in text.
Note: Maps not to scale.

exception to this is Figure 2.12). The pair at the bottom are also choropleth maps showing the distribution of Z-values for G_i^*. The lightest shade identifies areas that are clusters of low investment (Z < -1.26). The unshaded areas have no clusters. The darker shaded areas have high investment levels (Z > 1.56 for the darkest shade). Each map on the left is for the pre-reform period, each map on the right is for the post-reform period.

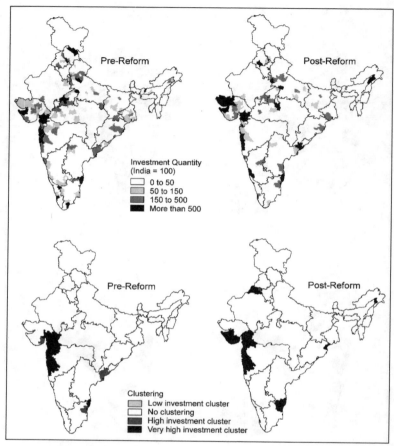

Figure 2.12: Distribution and clustering of industry in the chemicals sector

Source: Pre-reform data from ASI (data files for 1993–4); post-reform data compiled from CMIE (different years); authors' calculations. Details discussed in text.
Note: Maps not to scale.

Figures 2.7 and 2.8 show the spatial distribution of total investment in per capita and percentage share terms. In the pre-reform data, several pockets of high investment are seen: a long stretch from the Mumbai metropolitan region leapfrogging into southern and central Gujarat, in the Delhi metropolitan area stretching into northern Punjab, in the resource rich region of southern Bihar and north Orissa, and partially in the Calcutta metropolitan area. There was little or

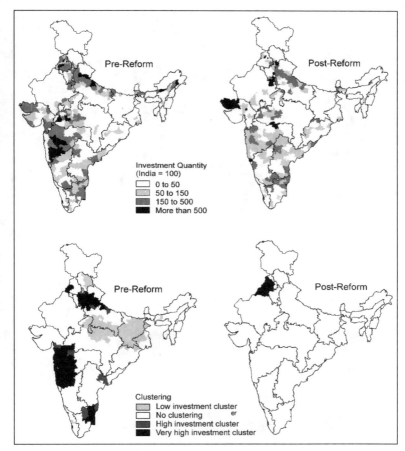

Figure 2.13: Distribution and clustering of industry in the agribusiness sector

Source: Pre-reform data from ASI (data files for 1993–4); post-reform data compiled from CMIE (different years); authors' calculations. Details discussed in text.
Note: Maps not to scale.

no investment in the far North-Eastern states, Jammu and Kashmir, northern Bihar and Uttar Pradesh, eastern Maharashtra and central Madhya Pradesh. The location of post-reform high investment pockets have changed in some cases: the earlier ones around Delhi, in Punjab, and in southern Bihar are considerably less intense, but the western pocket now stretches from Gujarat through the coastal districts of Maharashtra into Karnataka; Orissa now has a large high

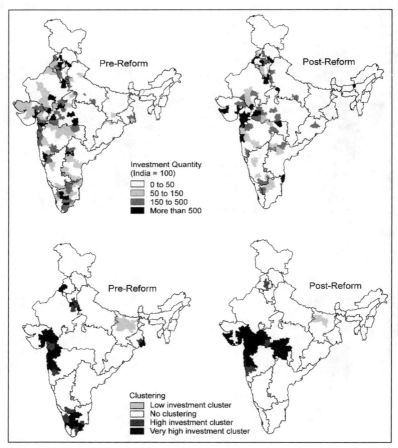

Figure 2.14: Distribution and clustering of industry in the textiles sector

Source: Pre-reform data from ASI (data files for 1993–4); post-reform data compiled from CMIE (different years); authors' calculations. Details discussed in text.
Note: Maps not to scale.

investment pocket in the north and another one shared with Andhra Pradesh, and a new swath can be seen in the Chennai region and south and west of it.

But a simple visual comparison of the pre- and post-reform maps does not help in identifying clear patterns. The G_i^* cluster maps of the same distributions (in the lower halves of the same figures), on the other hand, are quite compelling. In per capita terms, we can

PATTERNS OF INDUSTRIAL INVESTMENT 65

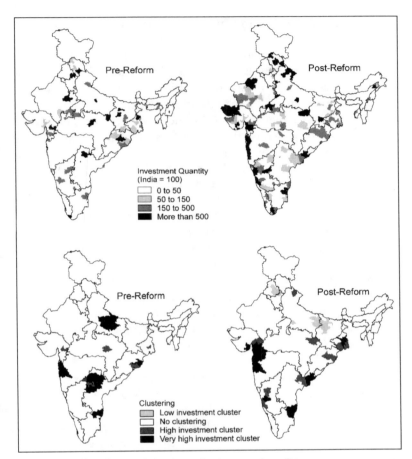

Figure 2.15: Distribution and clustering of industry in the utilities sector

Source: Pre-reform data from ASI (data files for 1993–4); post-reform data compiled from CMIE (different years); authors' calculations. Details discussed in text.
Note: Maps not to scale.

clearly see the spread of the large western high investment cluster into southern Gujarat, the relocation of the positive cluster in Punjab to the Himalayan foothills in Himachal Pradesh and Uttar Pradesh (now part of Uttaranchal), the decline of the eastern positive cluster, and the perseverance of the negative cluster in Bihar and eastern Uttar Pradesh. The clustered map of percentage share is similar, with the exception of the relocation effect in Punjab, and the stronger negative cluster effect in Bihar, Uttar Pradesh, and northern Bengal.

The maps disaggregated by capital source in Figures 2.9 and 2.10 confirm what we already expected. The distribution of private sector investment appears to mirror the distribution of total post-reform investment in Figure 2.7, an indication that the spatial pattern of new investments is being driven by the private sector. The private investments are focused on the western coast and Orissa, and widely spread over the south (except Kerala), especially the Chennai region and Vishakhapatnam, and north of the Delhi region. In general, there is less private investment in the north and east, especially Bihar, Rajasthan, eastern Uttar Pradesh, and West Bengal (note the strong negative cluster in Bihar). Central government investments, while patchier than private investments, appear to be more concentrated in the east (northern Orissa, southern Bengal, southeastern Bihar, eastern Assam), while bypassing large sections of the interior west and south. It is interesting to note that central government investment clusters appear in both the most advanced region (the west) and one lagging region (Bengal–Bihar–Orissa), but are not present in the rest of the country. Foreign direct investment is present in only a handful of districts (60 out of 470), with a clear coastal and metropolitan bias.

The sectorally disaggregated maps in Figures 2.11 through 2.15 are very informative. The most interesting maps are the ones for heavy industry (Figure 2.11), agribusiness (Figure 2.13), and textiles (Figure 2.14). The utilities maps (Figure 2.15) are interesting in only that they reveal the lack of pattern in either period; the chemicals maps (Figure 2.12) show little change in either format. The clustering effects in heavy industry have become decidedly stronger in the post-reform period; the high concentration patches in the east and west have become larger, and a new cluster has emerged around Chennai; the negative cluster in Bihar has spread wider and into parts of Uttar Pradesh and north Bengal. As indicated by the Moran's I statistics, the most remarkable change has taken place in agribusiness; the pre-reform pattern is very distinct—a very large positive cluster in Maharashtra, with smaller positive clusters in Tamil Nadu and Haryana / northern Uttar Pradesh, and a very large negative cluster stretching from central Bengal through Bihar into southern Uttar Pradesh and northern Madhya Pradesh. In the post-reform stage all these clusters have vanished, to be replaced by a solitary small positive cluster in Punjab. As noted before, the post-reform results for the agribusiness sector should be viewed with considerable scepticism. In the textile sector, the pre-reform stage had a large positive cluster in Maharashtra and Gujarat, smaller positive clusters in Tamil Nadu, southern Bengal, and around Delhi, and a negative cluster in Bihar. All that remains in the post-reform stage is the positive western cluster (now much larger) and the negative Bihar cluster (now smaller).

THE RETURN OF CUMULATIVE CAUSATION

Returning to the hypotheses outlined at the beginning of the chapter, consider what summary conclusion can be drawn at this point.

Hypothesis 1a: The advanced industrial regions will benefit more than the lagging un-industrialized regions. True, but with qualifications. It appears to be true that the advanced regions have been favoured by new investments, but within the advanced regions some profound changes are taking place: the old most-favoured districts no longer enjoy that primacy, but different districts, usually within the same clusters, have forged ahead; the sizes/shapes of the leading clusters have changed but the general locations of the clusters have not. It is tempting to look at Maharashtra's relative decline and declare that the advanced regions have fallen out of favour; but in reality, Maharashtra continues to be in favour, but the overbuilt, congested, expensive Greater Bombay district is relatively out of favour (but only as far as investment in manufacturing industry is concerned). Similarly, it may be tempting to look at the increase in Orissa's share of the new investments and conclude that lagging regions are rising now; but, thanks to government investment in heavy industry and utilities, even before the reforms, Orissa had had an investment level that was higher than the national average. After the reforms, Orissa's heavy industry sector has become stronger, but that may be a transient trend. Even if we claim that Orissa is one of the lagging states that has improved its position, the same cannot be said of the lagging giants Uttar Pradesh and Bihar, and to a lesser extent, West Bengal and Kerala. In Chapter 3, we will show that these states are shunned by private investment to an alarming degree; government investment does not avoid all these states like the private sector does, but the quantity and type of investment that the government is capable of making or willing to make cannot make up for the increasing shortfalls. In short, there is no doubt that the lagging regions have fallen further behind in the post-reform phase (as to what would have happened to them if there had been no reforms is an interesting question that cannot be answered here), and that advanced regions have forged further ahead (even while a significant internal reshuffling takes place within the advanced regions).

It is interesting to speculate on why these states—Uttar Pradesh, Bihar, West Bengal, and Kerala—continue to fare so poorly. Though at first glance these states appear to have little in common, one can argue that their local political economies are each distinguished by distributional struggles. One of the arguments in the 'new' growth literature (see the writings of Alesina and Rodrik 1994; Clarke 1995; Persson and Tabellini 1994; and others) is that there is a robust negative relationship between distributional struggle and growth performance. In Bihar and Uttar Pradesh, one sees evidence of bitter caste conflicts leading to identity-based politics and deeply factionalized political structures, so much so that some parts of these regions appear to be heading towards stateless anarchy. West Bengal and Kerala are stable, but they are both communist states which have historically placed egalitarianism and anti-capitalism at the top of their rhetorical and policy agendas. The local communist parties cater to their major constituencies—landless and small farmers, and organized industrial labour—and redistribution remains their winning issue. It is possible to argue that the prime reason for the failure of these states is their local political economy rather than some structural factor such as the lack of infrastructure or skilled labour. We will visit this issue of history and political economy in later chapters.

Hypothesis 1b: Metropolitan regions will be favoured over non-metropolitan regions. Superficially yes, but with a significant qualification. It is true that as in the pre-reform period, the metropolitan regions are likely to receive more investment than non-metropolitan regions, specially foreign and government, and to a lesser extent, private capital. But at the same time, the attractiveness of metropolitan regions as a whole is lower in the post-reform period than in the pre-reform period; indeed, the urban districts, in general, have performed remarkably poorly in the post-reform period. Remember that metropolitan districts as a whole have seen a decline in their national share from 23 per cent to 18 per cent; much of this decline is caused by Greater Bombay's loss of share, but the other metropolitan districts as a group have barely held their own. It is certain that metropolitan regions are no more attractive after the reforms than they were before the reforms; it is difficult to be more specific because one cannot strongly associate the beginning of reforms with the decline of metropolitan districts, which most likely was underway well before the reforms. Therefore, yes, metropolitan regions are more likely to benefit overall than non-metropolitan regions, but, more important, these benefits are not higher than they were before. Perhaps this is initial evidence of support for Krugman's thesis that liberalization breaks the monopoly power of metropolitan centres. More data are needed over a longer period, and especially after further liberalization, for a definite answer to emerge.

Hypothesis 1c: Coastal regions will be favoured over inland regions. Unequivocally, yes. Coastal districts have raised their share of national investment by over 10 percentage points (from under 35 per cent to over 45 per cent); this is despite the decline in Greater Bombay (which alone accounts for a 5.5 per cent loss nationally) and all of Kerala.

Hypothesis 2: The pattern of investment is going to differ by source of capital. Yes. The spatial pattern of domestic private sector investments, foreign direct investments, and central government investments are each clearly different. Foreign direct investments are coast and metropolis oriented, as theorized, and avoid the socialist states, but at merely 7 per cent of the total investment pool this is not yet a significant factor in the new economy. It is not possible to decipher any regional bias in this pattern. Domestic private sector investments favour the west and parts of the south and southeast, and are also coast oriented, but are less metropolis oriented. Central government investment patterns are the most inconsistent; there appears to be some bias toward the west coast and metropolitan areas, but, at the same time, of all the capital sources, this is the only one that has a significant presence in the east and in the interior of the country. In the next chapter, we model the determinants of central government and private sector investment location patterns and show that their decision-making imperatives are fundamentally different.

Is this a deliberate attempt by the state to mitigate some of the regional divergence caused by private investments? It is difficult to be sure, but it appears to be unlikely: first, much of the new central government investments are at locations where there already exist significant government investments in oil, coal, and steel (much of it in the east); second, the government appears to have no interest in staking out new industrial sectors for itself where it can invest in a new spatial pattern of its choice; third, the share of government investment in total investment is already shrinking; fourth, government (and other) investments in infrastructure, which is a large part of the government share, do not and cannot follow spatial patterns in the manner of other sectors like heavy industry or textiles. Therefore, even if the government were willing to allow regional equity principles to guide the location of new investments, it would be largely unable to do so. Therefore, though the spatial patterns of investment by the different actors are different, in the lagging regions where only the government has a significant presence, the problem of lack of investment cannot be solved by the central government. At best, it can maintain its current profile and stave off investment oblivion in the lagging regions, but any regeneration has to be led by the private sector.

Hypothesis 3: The metropolitan areas will undergo internal restructuring where the edges will rise over the core. Yes, with one exception. In Mumbai, Delhi, and Chennai, there is clear evidence that the suburban districts have become more important investment destinations than the core urban districts. We have referred to the Mumbai case several times, but even more dramatic turnarounds have taken place in Delhi and Chennai. Calcutta is the only exception, but the quantity of investment in that metropolis (city or suburban district) is so low that it should not be considered a serious deviation from our theoretical expectations.

The findings provide significant if not conclusive evidence of support for the theoretical approach outlined in Chapter 1, that is 'cumulative causation' and 'increasing returns' where the leading regions (defined in meta, meso, or micro terms) keep advancing more rapidly than lagging regions (some of which may not be advancing at all). There are minor exceptions, some reorganization within the leading group, and evidence of some spread of industrial activity, but the basic picture is quite clear. The emerging spatial pattern is led by investments by the private sector which is demonstrably averse to lagging regions, just as the central government is becoming a weaker player. This brings up the second observation on the development prospects of the lagging regions. If the state will not or cannot be any more involved (it can only be less involved in the foreseeable future), and the private sector cannot be induced till some local political-economic problems are resolved, and these local problems may not be resolved without investment and growth, how can development reach the lagging regions, and in the long run, how can the reforms be sustained? If the pattern of coastal bias continues regardless of the resolution of the local political-economic situations, is the development of the interior possible without state intervention? In short: what happens to Bihar? We return to this question in the final chapter.

CONCENTRATED DECENTRALIZATION

Can the Indian experience on industrial location be placed in the context of other developing nation experiences? There is evidence from South Korea (Henderson 1988; Lee 1987; World Bank 1999) that as a result of liberalization in the 1980s, the reinstatement of local government autonomy, and heavy communication infrastructure investments by the state away from Seoul and Pusan, manufacturing industry has decentralized. Among other physically small countries, there is a clear trend toward regional de-concentration of income and industry in Colombia (see Cárdenas and Pontön [1995] on regional convergence and Lee [1987, 1989] on the de-concentration of employment at the metropolitan level). In Indonesia, a larger country, the summary data suggest that some industrial de-concentration has taken place during the reform period (the early 1980s onwards) in the secondary and tertiary sectors of the economy, while government consumption expenditure and fixed capital formation have become more concentrated (Akita and Lukman 1995); others note that 'these aggregate groupings...conceal important trends at the sub-regional level.... There has almost certainly been a rising concentration of industrial activity on the fringes of major urban concentrations, such as Jakarta and Surabaya' (Aswicahyono, Bird and Hill 1996: 356; also see Henderson and Kuncoro 1996). Among countries that may be considered geographically comparable to India, the industrial location experience of Brazil has been widely studied. Townroe and Keen (1984) and Diniz (1994) have argued that 'polarisation reversal' has taken place in the state of São Paulo in Brazil, where an 'agglomerative field' with a radius of around 150 kilometres around São Paulo is seeing faster industrial growth than the city itself.[8] According to Diniz, however, inter-regional industrial polarisation into the Southeast and the state of São Paulo continues.

Similarly, it appears that what is happening in India is quite distinct from 'spread' or 'polarization reversal' in that there is increasing inter-regional polarization of industry at the same time that there is intra-regional dispersal in the leading regions. In other words, the situation is one of concentration with dispersal, or to make an ironic use of the term, of concentrated decentralization, where the new growth centres are in the advanced regions rather than in the lagging ones.[9] Though the reforms of 1991 were used as the point of departure in the analysis, it does not follow that this concentrated decentralization process is purely the result of the reforms. First, according to some scholars, the reforms began with Rajiv Gandhi in 1985; second, it is quite likely that some elements of the process, especially that of the decline of metropolises as destinations of industrial investment, were well under way before any such formal, sharp transition (in 1985 or 1991). At this point, we are unable to discern whether the changes are driven by changes in transport costs, agglomeration economies, or political action. We suggest that this is an issue of secondary importance, that is, relevant only at the intraregional scale. Remember that large parts of the country are marked by what can be termed 'intangibles': Gujarat's entrepreneurial culture, virtual statelessness in Bihar, extreme labour militancy in West Bengal, knowledge-based cultures in the south, etc.

It is possible that these elements of local political economy create the primary foundation on which inter-regional outcomes turn. That is, the industrial location decision is a two step process: First comes the decision on which general region to invest in, followed by the

PATTERNS OF INDUSTRIAL INVESTMENT 73

decision on the specific location within the selected region. The first decision leads to inter-regional divergence; the second decision may lead to intraregional convergence (due to declining transport costs and/or rising agglomeration diseconomies). In India, the key is to devise strategies that will influence the first decision.

NOTES

1. After the data were collected, the states of Uttar Pradesh, Bihar, and Madhya Pradesh were reorganized and three new states (Chhattisgarh, Jharkhand, and Uttaranchal) were carved out of them. We do not use these new state definitions in our presentation, largely because it is difficult to compare pre- and post-reform conditions using the new boundaries.
2. The CMIE data are likely to have some problems. Primarily, it is possible that the CMIE's method of data collection has introduced some bias; their lists were compiled from official announcements (on the granting of Letters of Intent, or the signing of Memoranda of Understanding, etc), press reports, and company announcements and reports. This is unlikely to generate a complete list. Second, since the CMIE is Mumbai-based, its information on the west may be better than for other zones (especially the farthest zone, the east). As a result, their lists may be biased toward showing relatively more investment in the west. But it is expected that both problems are likely to be true for small projects (perhaps Rs 500 million or less), as bigger projects will have generated so much publicity through various media that they are likely to have been included in the CMIE lists.
3. The share of total investment allocated to each sector is reasonably consistent. Gains have been made in the heavy industries sector, where the pre- and post-reform shares were respectively 33.4 and 37.2 per cent, and in the chemicals and petroleum sector, where the shares are 18.4 and 26.2 per cent; losses have been in the utilities sector with pre- and post-reform shares of 30.3 and 25.2 per cent, in textiles with 7.8 and 5.7 per cent, and in agribusiness with 7.1 and 2.9 per cent. We will suggest later that the CMIE agribusiness data are unreliable.
4. These data do not include figures for Jammu and Kashmir and the far North-Eastern states (Arunachal Pradesh, Manipur, Meghalaya, Mizoram, Nagaland, and Sikkim). Given the turmoil in Jammu and Kashmir, its data are unlikely to be reliable. In addition, all these states together accounted for less than 0.4 per cent of the total new industrial investment, almost entirely in utilities projects (typically hydel power), typically spread over several districts. Also note that the maps show data at the district level, but in order to minimize visual clutter, the district boundaries have not been drawn.
5. Three hundred and fifty-seven of the 470 odd districts in India received some investment. 54 districts are classified as coastal districts and 26 are

classified as metropolitan districts. The identification of coastal districts is straightforward: any district on the Arabian Sea or Bay of Bengal has received this designation. We have used standard definitions of 'greater' or 'metropolitan area' for defining the metropolitan areas of the following cities: Mumbai, Calcutta, Delhi, Chennai, Bangalore, Hyderabad, and Ahmedabad. For instance, the Calcutta metropolitan area includes the districts of Calcutta, Haora, Hugli, North 24 Parganas, and South 24 Parganas. Ahmedabad is the only city for which no additional (suburban) district has been added. See Figure 2.1 where the four major cities and their suburban districts have been identified.

6. It is important to note that the relative post-reform government shares may look different if the local government data (especially in the power sector) are included. Though there is anecdotal evidence that this involvement has diminished after the reforms, we do not have specific data.

7. This may indicate that the post-reform or CMIE data for the agribusiness sector are incomplete. In the pre-reform period, the agribusiness sector had the highest level of clustering. It is difficult to imagine a scenario whereby this extremely clustered industry would suddenly lose the circumstances which enabled clustering in the first place, unless, as expected, there are missing data. Remember the caution about missing small projects in the CMIE data. Since the agribusiness sector is most likely to be composed of small factories (in the ASI data, this sector includes about 23 per cent of all factories but only about 8 per cent of invested capital and 5 per cent of fixed capital), it is likely that the CMIE data are incomplete for this sector. Henceforth, the results for the agribusiness sector should be considered suspect.

8. The idea of 'polarization reversal' followed from Hirschman's notion of polarized development. This theory has been investigated by Richardson (1980) and Chakravorty (1994). The latter has noted that the meaning of polarization reversal depends on the definition of 'pole'. He argues that declines in inter-urban, intra-metropolitan, and intra-regional concentration are all possible and likely in most developing nations, but inter-regional inequality and concentration are likely to endure in large countries.

9. The term 'concentrated decentralization' has been used in the growth pole literature, where it meant that rather than scatter scarce investment resources around equally, 'small centers or cores are set up in the periphery, thus to some extent distributing investment but also taking some advantage of urbanisation economies' (Darwent 1975: 553). For about two decades (the 1960s and 1970s) the growth pole policy was considered a critical tool that could help realize agglomeration efficiencies while advancing the cause of inter-regional equity, making it one of the most discussed ideas in urban and regional development (Hansen 1967; Lo and Salih 1981).

3
Determinants of Industrial Location

In order to understand the prospects of comparative regional development it is necessary to understand the logic behind the location of new investments. We know that the geographical unevenness of international investment flows rivals, even exceeds, the unevenness of world incomes and development levels (Amirahmadi and Wu 1994; de Mello 1997). There is also clear evidence that both within nations that are able to attract FDI and nations that are not, uneven regional development is increasingly becoming a significant issue (see Fan 1995 and Wei 2000 on China; Akita and Lukman 1995 and Aswicahyono, Bird, and Hill 1996 on Indonesia; Ahuja *et al.* 1997 and Daniere 1996 on Thailand, etc.). This uneven development arises largely from the geographical unevenness of the distribution of new investment capital. There are several reasons, arising primarily from factors that dictate industrial location decisions, that explain why there is unevenness in the geographical distribution of industrial capital.

In this chapter, we identify these factors and show how the increasing importance of private capital is shaping the new geography of unevenness. State capitalism, which in the pre-reform period followed a location logic that was not entirely driven by efficiency, is in retreat; as a result, now that the private sector is the source of the bulk of the new industrial capital, the investment gaps between rich and poor regions in terms of industry and infrastructure is being further widened. In other words, because the source of the majority of new industrial capital is different (that is, from the private sector), increasing regional inequality is the expected outcome of liberalizing structural reforms.

We begin from the question: what factors influence industrial location? From Chapter 2 we already know where new industrial investments locate both before and after structural reforms. Following the basic arguments outlined earlier—that the reforms of 1991 can be seen as a watershed for the political economy and the economic geography of growth in India—we now undertake the analysis in two steps. In the first step we create a model of industry location based on factors, that is, capital, labour, infrastructure, regulation, and geography, that have been identified in the literature. We estimate this model with district level data for all industry sectors and for selected sub-sectors identified in Chapter 2 (heavy industry, chemicals, textiles, agribusiness, and utilities). The results show that industry location follows cumulative causation principles—existing industrial districts and their proximate districts are the primary beneficiaries of new industrial investment. Following this, we use the same model to show that private and state capital have fundamentally different determinants of location, and that private capital seeks locations that create more spatial inequality.

THE GENERAL FACTORS

It is necessary to begin by identifying the factors that influence industry location. Let us assume that all location decisions are made in the private sector (later we will show how the same model can be used for comparison with the public sector). That is, let us presume that market considerations are the only ones that need to be factored into the industrial location decision. In the empirical literature, there are two broad approaches to identifying the factors that influence firm location: One is a survey approach that asks location decision makers what is important to them. The other is a modelling approach to try to identify the revealed preferences based on site/region characteristics. A large number of factors, with some overlap, have been identified using these two approaches (see Chapman and Walker 1991 for an extensive list). In theory, the most important firm location criteria are market access, infrastructure availability, agglomeration economies (localization economies arising from knowledge spillovers and intra-industry linkages, access to specialized labour pools, and upstream and downstream process integration, and urbanization economies such as access to specialized services, a diverse labour pool, inter-industry information transfers, general social infrastructure and urban amenities), and state regulations on environmental and pollution standards, incentives in lagging regions or for emerging technologies, and the general level of political support (see Hanushek and Song 1978; McCann 1998; Webber 1984;). Not surprisingly, the survey-based approaches reveal that there is a substantial random element in the choice of location: Personal reasons, chance, and opportunity (such as finding a good site) are given as explanations almost half the time (see Mueller and Morgan 1962; Calzonetti and Walker 1991).

Since we have not asked the decision makers about their choices, let us use the second or revealed preference modelling approach. The factors identified in the literature in the foregoing discussion may be organized into the following categories:

(1) *Capital,* which refers to the quantity and productivity of the existing capital investments, and the availability of industrial capital from lenders. Note that the quantity of capital in the same industry is a measure of local concentration and a measure of localization economies—it suggests the availability of specialized labour pools and buyer–supplier networks. High capital productivity (or output per unit of invested capital) and the availability of capital for borrowing are indicators of efficiency in the given industry.

(2) *Labour,* which refers to the size of the industrial and total labour pool in the region, and the productivity of industrial labour. The size of the industrial labour pool is a measure of urbanization economies—it suggests the presence of thick labour markets and inter-industry labour transfers. The total labour pool or population is also a measure of market size; the larger the population the greater the market access. High levels of labour productivity (or output per unit of labour) are indicative of capital intensity, which signals the existence of large plants with significant internal scale economies and/or high technology firms in the same industry.

(3) *Infrastructure* includes elements of physical and social infrastructure. Physical infrastructure such as roads and transportation hubs (ports, airports) are widely considered to be key determinants of plant location. Indicators of social infrastructure such as health and education standards provide an understanding of the quality of life conditions, and may be considered to be amenities for workers, which may be critical for some industries.

(4) *Regulation* broadly refers to the system of incentives (such as tax breaks) and disincentives (such as environmental standards) which have to be factored into the location decision. It is difficult to get this kind of highly localized or disaggregated information for the whole nation. This is specially true of India where the key to decision-making may not be the localized incentive system but a sense of political support for private sector led industrialization in the region. Regimes that are ideologically opposed to liberalization are unlikely to provide the conditions that welcome new private investments, or may be perceived to be unfriendly to capital.

(5) *Geography* includes spatial characteristics such as coastal or metropolitan location. Coastal locations provide access to the external world (which may be very important in export-oriented regimes) and physical amenities desired by high level managers. Metropolitan locations provide large local markets, urbanization economies, and, often, localization economies.

PRIVATE AND PUBLIC CAPITAL

Now we turn to the hypothesis that the location logic of state capital is different from that of private capital. Private capital seeks profit maximizing or efficient locations, which are the already leading industrial regions that provide the necessary infrastructure and economies of agglomeration. On the other hand, the location decisions of state capital are not as oriented toward the leading industrial regions because these decisions are also based on equity and security considerations. We argue that given the increasing dominance of private capital over state capital in liberalized economies, the resulting rise in regional inequality will have to be mitigated by state action.

The theoretical literature on industrial location, and location theory in general, began with Weber (1909) and continued through the writings of Hotelling (1929), Hoover (1948), Isard (1956), Alonso (1964), and others. This literature, much of which we have already reviewed earlier, has sought to explain patterns of urban land use and the optimal location of profit maximizing firms in a space economy with varying costs of land, labour, and transportation inputs. Later, the thrust of the discussion shifted to firm location within urban systems; this large literature focused on the effects of the centripetal and centrifugal forces of agglomeration economies and diseconomies (Richardson 1973; Henderson 1988). More recently Paul Krugman and his associates (Krugman 1996; Fujita, Krugman, and Venables 1999) have argued that firm location decisions are based primarily on transportation cost—industries are geographically dispersed when transport costs are high, concentrated when costs decline to some point, and dispersed again when the costs decline even further. We will examine these theses in detail in Chapter 4.

This entire literature presumes that all capital is private capital, and all location decisions are made by profit maximizing private firms. The presence of the state as a significant entity that owns and locates firms and industries is not considered. There are at least three reasons why this is an omission of some consequence:

(1) First, state decisions on industry location are not necessarily or usually profit maximizing (more on this in the next paragraph).
(2) Second, in all developing and ex-socialist nations, industrialization has been state-led, so that the state, to some degree, still owns the 'commanding heights' of the industrial sector.
(3) Third, state industrial location decisions have considerable influence on the location decisions of private firms (mainly through the provision of shared infrastructure and localization economies).

More descriptive approaches in the geography and regional science literatures do recognize the 'different locational considerations' of state capital, especially the following ones: first, the need to include and provide for 'the poor and the geographically peripheral'; second, the absence of competition in what are often (loss-making) monopolies; third, the need to seek popular support, and the use of state investment as a method of doing so; fourth, the use of industrial location as the principal tool in regional development policy.

These arguments, with reference mainly to infrastructure provision in developed economies, are summarized in Harrington and Warf (1995). In addition, one must consider the location of security oriented or defence related industry which is obviously not dictated by market factors.[1] Yet, despite what appears to be an unambiguous situation—that state capital follows different (non-market) location considerations relative to private capital—there is virtually no empirical evidence from developing nations on the extent to which this is true, nor is there evidence on the extent to which different industrial location factors have varying influence on the behaviour of state versus private capital.[2]

A MODEL OF NEW INDUSTRY LOCATION

Let us now identify the specific factors that contribute to the location decision of new industry. In earlier chapters, we have suggested that concentration processes are important in the location decision. This implies that new industrial investments are likely to be located near existing industrial investments. Not only that, but new investments are likely to be near locations where other new investments are being made, which means that new investments are also likely to be clustered. That is, concentration and clustering are expressed through investments: existing or old investments, and new investments in proximate or surrounding areas respectively. In addition, some locations have advantages deriving from capital availability and capital productivity, labour availability, labour skills and labour productivity, physical and social infrastructure, political support, and spatial phenomena such as access to consumer markets and coastal regions. Following the earlier discussions, a general model of new investment location determination can be written formally as

$$I_{new} = C + K + L + I + R + S + e \qquad (1)$$

where C (the constant term) and e (the error term) are interpreted in the usual way, and K, L, I, R, and S represent sets of explanatory Capital, Labour, Infrastructure, Regulation, and Spatial/Geographical variables respectively.

I_{new} is the log transformation of the raw investment amount, where the investment amount depends on the sector being modelled. That is, I_{new} is $I_{new\text{-}chemicals}$ when only the chemicals sector is being considered; similarly, in the second part of the analysis, I_{new} is $I_{new\text{-}private}$ when only private sector investments are considered, and is $I_{new\text{-}government}$ when only central government investments are considered.

DETERMINANTS OF INDUSTRIAL LOCATION   81

K represents a set of three Capital variables:[3]

OLDINV is the log transformation of the total pre-reform investment. OLDINV is expected to be positively related to I_{new}, especially to $I_{new\text{-}private}$. That is, new investments, especially new private sector investments are expected to favour districts with existing industrial investments.

IND-CREDIT or industrial credit, is given by the per capita lending to local industry by financial institutions, defined as the per capita bank credit to industries derived from the information on scheduled commercial bank branches, deposits and credits on the last Friday of March 1993. *Source*: CMIE 1993.

PROD-CAPITAL is a measure of the productivity of capital at the district level, and is calculated as the value added per unit of fixed capital for existing industry. *Source*: calculated from the ASI data.

L represents a set of three Labour variables:

POP-DISTRICT is the log of district population. *Source*: CMIE 1993 from the 1991 Census.

LABOUR-INDUSTRY is the percentage of workers employed in non-household manufacturing industry. *Source*: CMIE 1993 from the 1991 Census.

PROD-LABOUR is a measure of the productivity of labour and is calculated as the value added at the district level per unit of factory labour. *Source*: calculated from the ASI data.

I represents a set of three Infrastructure variables:

INFRASTRUCTURE or physical infrastructure, is a measure of access to physical infrastructure, and is calculated as a function of proximity to national highways, airports and ports. The values of INFRASTRUCTURE range from 0 to 3, where 3 represents a situation where the given district has at least one national highway passing through it (weight 1), has at least one airport within 100 kilometres (weight 1), and has at least one port within 100 kilometres (weight 1); correspondingly, 0 implies that the given district has no national highway passing through it, nor an airport or port within 100 kilometres. *Source*: calculated using Geographic Information Systems (GIS) on government of India provided base maps of highways, airports, and ports.

LITERACY is the percentage of the adult population that is literate by government of India standards. *Source*: CMIE 1993 from the 1991 Census.

INFANT-MORT, is the mortality rate at age five years per 1000 live births, estimated from the 1991 Population Census. *Source*: Rajan and Mohanchandran 1998.

The only Political variable P is:

SOCIALIST, which is a dummy variable that takes a value of 1 for every district in West Bengal and Kerala. Districts in Tripura (another socialist state) were not used in the analysis, and we chose not to assign districts in Bihar as socialist. Bihar has what may be called a populist caste-based government, and giving it the distinction of socialism, for better or worse, may be inappropriate. The other problem with including Bihar in this category is that every other state that has had left-of-centre governments in the early 1990s (such as Karnataka and Orissa) would also have to be similarly characterized. As far as this variable is meant to represent political will, which may be resistance to liberalization, or its counterpart, enthusiasm for reforms, Bihar should be so categorized. But, understanding the lack of investment in Bihar is an important goal, and we have preferred not to cloud the issue by introducing the socialist element.

S represents a set of two Spatial variables:

COASTAL, a dummy variable that takes a value of 1 for all coastal districts (57 districts were classified coastal). The identification of coastal districts is straightforward: Any district on the Arabian Sea or Bay of Bengal has received this designation.

METROPOLITAN, a dummy variable that takes a value of 1 for all metropolitan districts, that is, the core city district and the surrounding suburban districts (26 districts were classified metropolitan). We have used standard definitions of 'greater' or 'metropolitan area' for defining the metropolitan areas of the following cities: Mumbai, Calcutta, Delhi, Chennai, Bangalore, Hyderabad, and Ahmedabad. For instance, the Calcutta metropolitan area includes the districts of Calcutta, Haora, Hugli, North 24 Parganas, and South 24 Parganas. Ahmedabad is the only city for which no additional (suburban) district has been added.

SPATIAL-LAG is a term that corrects for spatial autocorrelation (see below) and also has geographical meaning. It is a measure of spatial clustering, and the parameter estimates for this term will indicate the degree to which new investments cluster together; for instance, the extent to which $I_{newPrivate}$ is likely to locate in the proximity of other $I_{newPrivate}$.

NOTES ON METHODOLOGY

An explanation of the spatial lag term is useful at this point. As discussed in Chapter 2, spatial autocorrelation (where similar values cluster in space: that is, high values are proximate to high values, and low values are proximate to low values) is a common feature of spatial distributions. There are several tests of spatial autocorrelation, the most well known of which is Moran's I. We showed in Chapter 2 that industrial investments in India demonstrate very strong spatial autocorrelation, that is, high investment districts were found to cluster, as were low investment districts. The existence of spatial autocorrelation or spatial dependence poses serious problems in regression modelling, much like serial autocorrelation does (see Anselin 1995). One of the ways of dealing with this problem in a regression modelling format is to add a 'spatial lag' term on the right hand side. Following Anselin (1992: 18–1) 'spatial lag is a weighted average of the values in locations neighboring each observation…if an observation on variable x at location i is represented by x_i, then its spatial lag is $\Sigma_j w_{ij} x_j$,' which is the sum of the product of each observation in the data set with its corresponding spatial weight.

There are several possible spatial weights: based on contiguity (considering only adjacent parcels) or distance. We used a number of spatial weights for analysis, but the one presented here is the distance weight, where the cut-off distance is 150 kilometres. That is, for a given district i, every district whose centroid lies within 150 kilometres of the centroid of i was considered a neighbour; the sum of the I_{new} of these neighbours is the spatial lag for district i. Let us say that a given district has six districts which lie within 150 kilometres of its centroid. The total investment in those six districts would be the spatial lag of the given district. Hence in a high investment cluster, all districts would have high values of spatial lag. The argument for using the spatial lag correction for a given district in that cluster is that its high investment is not independently caused by the regressors, but is dependent on its regional feature of high investment. Alternatively, a district located in a low investment cluster is also likely to receive low investment.[4] Therefore the spatial lag term corrects for spatial autocorrelation in spatial regression models, and, just as important, at the same time it is a measure of clustering.

The problem of spatial autocorrelation (which is relatively easily fixed) is only one of the problems of modelling this data set. Another serious problem arises, if one is to consider OLS models, because the assumption of normality of the dependent variable is seriously violated. Depending on the form of I_{new} to be used (that is, for all industry, or heavy industries, or chemicals, or the private sector, etc.) there are large numbers of districts that have no investment; note that these are not missing data, but are real measured absence of investment. The following list shows how many of these districts had non-zero investment:

All industry: 327 districts with investment, 78 without investment;
Heavy industries: 243 districts with investment, 162 without investment;
Chemicals: 137 districts with investment, 268 without investment;
Textiles: 118 districts with investment, 287 without investment;
Agribusiness: 194 districts with investment, 211 without investment;
Utilities: 179 districts with investment, 226 without investment;
Private sector: 292 districts with investment, 113 without investment;
Public sector: 164 districts with investment; 241 without investment.

Hence, it is not possible to use OLS models on the full data set. But using only the non-zero data does not allow analysis of the absence of investment. Also, there is the possibility that the estimates for the non-zero investment would be biased. The solution is to use two sets of models—a logistic model set where the complete data are analysed to estimate the probability of a district receiving some investment; and an OLS/Heckman selection model set to estimate the contribution of the explanatory variables to the quantity of investment.

In the logistic models the dependent variable takes a value of 1 when there is some non-zero investment (call this 'success'), and 0 when there is no investment (call this situation 'failure'). The Wald statistic is commonly used for significance testing of the parameter estimates in logistic models. When the parameter estimate is given by ß and its asymptotic standard error given by ASE, the Wald statistic for the two-sided alternative is given by $[ß/ASE]^2$ (See Agresti 1996 on the design and interpretation of logistic models).

The problem of estimating the determinants of the quantity of new investment is investigated using the Heckman selection modelling method. This is a two step method that uses the complete data. It presumes that the zero values for the dependent variable are actually missing values; that is, the sample is presumed to be non-randomly selected. The Heckman procedure first runs a probit model on the full data, estimates the values of a 'missing explanatory variable' resulting from the possible selection bias, and then estimates the determinants of the non-zero values using an OLS model. We have run both OLS and Heckman selection models with the data for industry sectors (the Heckman models were not run for the Private/Public models). We have reported the findings from the Heckman selection model only in the two cases (all industry and agribusiness) with significant selection bias (indicated by the estimates of lambda, a model statistic reported for all the models). We have also reported the OLS generated adjusted R-square values for all models to indicate their predictive ability; hence, in the two cases where the Heckman selection estimates are reported, the adjusted R-square figures are from the corresponding OLS model. It is to be noted that the Heckman selection parameter estimates, even in the two cases where there is selection bias, are virtually identical to the OLS estimates. Good surveys of the Heckman two stage estimation process are in Vella (1998) and Winship and Mare (1992).

Table 3.1: Logistic models of new investment
(N = 405)

Variable	Total Investment	Heavy Industry	Chemicals	Textiles	Agribusiness	Utilities
OLDINV	0.121	.0135	0.192	0.203	0.138	0.051
	(6.44)**	(22.54)***	(37.74)***	(24.07)***	(20.93)***	(2.74)*
NEWINV-LAG	0.311	0.422	0.305	0.748	0.532	0.207
	(9.00)***	(22.58)***	(6.11)***	(28.99)***	(17.95)***	(4.27)**
IND-CREDIT	0.0003	0.0007	0.0003	0.0003	0.0003	0.00005
	(0.21)	(2.29)	(1.51)	(1.57)	(1.21)	(0.05)
PROD-CAPITAL	-0.455	-0.007	-0.420	-0.252	-0.283	0.002
	(2.37)	(0.14)	(1.60)	(0.71)	(0.98)	(0.05)
POP-DISTRICT	0.980	0.802	0.405	0.243	0.567	0.778
	(14.29)***	(13.17)***	(2.79)*	(0.87)	(7.05)***	(13.52)***
LABOUR-INDUSTRY	0.082	0.079	0.623	0.060	0.072	0.075
	(1.53)	(2.71)*	(2.55)	(2.23)	(4.04)**	(4.47)**
PROD-LABOUR	0.006	0.0004	0.001	0.004	0.0006	0.001
	(8.76)***	(0.18)	(4.15)**	(2.66)	(0.78)	(7.12)***
INFRASTRUCTURE	0.498	0.014	-0.082	0.364	0.171	0.335
	(6.59)**	(0.01)	(0.24)	(4.50)**	(1.47)	(7.38)***
LITERACY	0.002	0.009	-0.277	0.0004	-0.018	0.015
	(0.02)	(0.44)	(3.35)*	(0.00)	(1.99)	(1.57)
INFANT-MORT	-0.004	0.003	0.003	0.0004	0.002	0.001
	(0.88)	(0.49)	(0.42)	(0.01)	(0.16)	(010)
SOCIALIST	-1.001	-0.933	-1.02	-1.178	-1.308	-1.229
	(2.09)	(2.00)	(2.56)	(2.84)*	(4.41)**	(4.98)**
COASTAL	-0.791	0.480	1.136	-0.491	0.939	0.164
	(1.19)	(0.74)	(5.03)**	(1.23)	(4.20)**	(0.15)

Table 3.1 (contd)

Variable	Total Investment	Heavy Industry	Chemicals	Textiles	Agribusiness	Utilities
METROPOLITAN	3.566	4.20	1.484	0.742	1.365	0.710
	(0.08)	(0.20)	(1.70)	(0.73)	(1.37)	(0.68)
Constant	-9.135	-9.366	-5.457	-6.663	-6.473	-8.296
	(14.77)	(20.38)***	(6.31)**	(7.34)***	(11.45)***	(18.84)***
Correct prediction (%)	86.17	80.74	83.46	85.19	75.06	73.33
Chi-square	141.87	207.98	197.51	213.75	164.59	112.99

Source: Authors' calculation.
Notes: Figures in parentheses are Wald statistics.
*** Significant at 1 per cent; ** Significant at 5 per cent; * Significant at 10 per cent

Table 3.2: OLS and Heckman selection models of new investment

Model Variable	Heckman Total Investment (N = 347)	OLS Heavy Industry (N = 243)	OLS Chemicals (N = 137)	OLS Textiles (N = 118)	Heckman Agribusiness (N = 194)	OLS Utilities (N = 179)
OLDINV	0.198***	0.093***	0.026	0.059	0.119***	0.078**
	(4.32)	(2.92)	(0.60)	(1.41)	(4.73)	(2.07)
NEWINV-LAG	0.103	0.163**	0.029	0.111	0.545***	-0.059
	(1.59)	(2.03)	(0.21)	(0.89)	(5.68)	(0.52)
IND-CREDIT	0.00009	0.00001	0.0002	0.00009	0.0001	0.0002
	(0.86)	(0.09)	(1.62)	(0.85)	(2.03)**	(1.48)
PROD-CAPITAL	-0.057	-0.043	-0.666*	-1.625***	-0.132	0.0005
	(0.23)	(0.52)	(1.86)	(2.97)	(0.42)	(0.18)
POP-DISTRICT	0.171	0.407*	-0.211	0.002	0.115	-0.661**
	(0.96)	(1.92)	(0.65)	(0.01)	(0.86)	(2.48)
LABOUR-INDUSTRY	0.040*	0.011	0.010	0.055*	-0.009	0.008
	(1.67)	(0.38)	(0.26)	(1.72)	(0.58)	(0.25)
PROD-LABOUR	0.0014***	0.0006	0.0006**	0.003*	0.00001	0.00008
	(3.06)	(0.91)	(2.01)	(1.93)	(0.04)	(0.52)
INFRASTRUCTURE	0.057	0.020	0.040	-0.008	-0.025	0.058
	(0.56)	(0.15)	(0.19)	(0.06)	(0.29)	(0.40)
LITERACY	0.0009	-0.011	0.004	-0.017	-0.005	-0.032**
	(0.10)	(0.98)	(0.22)	(1.21)	(0.68)	(2.32)
INFANT-MORT	0.003	0.001	0.005	-0.001	0.003	0.0001
	(1.11)	(0.42)	(0.98)	(0.28)	(1.22)	(0.03)
SOCIALIST	-0.763*	-0.875*	-0.818	-0.837	0.445	1.227**
	(1.93)	(1.71)	(1.09)	(1.22)	(1.14)	(2.10)

Table 3.2 (contd)

Model Variable	Heckman Total Investment (N = 347)	OLS Heavy Industry (N = 243)	OLS Chemicals (N = 137)	OLS Textiles (N = 118)	Heckman Agribusiness (N = 194)	OLS Utilities (N = 179)
COASTAL	1.02***	0.507	1.856***	0.169	−0.062	1.899***
	(3.38)	(1.44)	(3.74)	(0.44)	(0.30)	(4.77)
METROPOLITAN	0.510	0.508	−0.305	0.339	0.021	0.971
	(1.01)	(0.92)	(.046)	(0.65)	(0.06)	(1.53)
Constant	0.286	0.903	4.978*	4.384**	−0.467	11.013***
	(0.18)	(0.51)	(1.78)	(2.10)	(0.37)	(4.66)
Model Wald chi² (Heckman Model)	111.03***	35.54***	25.38**	31.19***	63.77***	43.07***
Heckman Model lambda (std. err.)	0.674 (0.349)	0.287 (0.437)	−0.358 (0.886)	−0.039 (0.746)	1.09 (0.186)	0.271 (0.867)
Adjusted R-Sq. (Linear Model)	0.252	0.134	0.157	0.146	0.112	0.151

Source: Authors' calculations

Notes: *** Significant at 1 per cent; ** Significant at 5 per cent; * Significant at 10 per cent. Figures in parenthesis are Z scores for the Heckman models, t statistics for the OLS models. The Adjusted R-squares reported for the Heckman selection models are from the OLS models with the same functional forms.

MODEL FINDINGS FOR INDUSTRIAL SECTORS

Let us begin with a discussion of the findings for the different industry sectors (including all industry). The findings on the Private/Public differences are discussed in a separate section below. The logistic model results are reported in Table 3.1 and the OLS/Heckman selection model results are reported in Table 3.2. Recall that the logistic models are to be used to identify the factors that contribute to a district getting some (as opposed to no) investment, and the OLS/Heckman selection models are to be used to identify the factors that contribute to the quantity of new investments. In the following discussion, the results in the two tables are considered together and focused around the regressors or sets of independent variables.

But first consider the model sets as a whole. The logistic models are generally robust and predict over 73 per cent of the distribution of districts with zero and non-zero investments. The OLS/Heckman selection models are robust, but generally less successful in explaining the variation in the distribution of new investments. The adjusted R-square values go from a low of 0.112 for agribusiness to a high of 0.252 for all industry. It is possible to improve the R-square values by adding variables which are known to be influential; for instance, adding dummy regional or state variables should show the attractiveness of the western region relative to the eastern region; the addition of a second spatial lag term, this time for the clustering effects of existing (or pre-reform) investment, is also likely to improve the model fit. But these adjustments would not add to our understanding of the investment distribution process. There are surely local factors we have not inserted in the models that have a significant bearing on location decisions of individual firms. This is an important point which we will discuss later.

The concentration variable—old investment (OLDINV)—is consistently significant in all logistic models, and in all but chemicals and textiles among the OLS/Heckman selection models. The clustering variable—new investment in neighbouring districts (SPATIAL-LAG)—is also very strongly significant in all the logistic models, but not significant for all industry, chemicals, textiles, or utilities in the OLS/Heckman selection model set. The parameter estimates for SPATIAL-LAG are higher, indicating that the quantity of new neighbouring investment has a stronger influence on the existence and quantity of new investment in a district. Generally, we can conclude that concentration and clustering processes, expressed through existing and new nearby investments, are very important in determining whether or not a district gets any new investment, and are important, but less so, in determining the quantity of the new investment.

The other capital variables—availability of industrial credit and the productivity of capital—have virtually no influence on the location and quantity of new industrial investments. In fact, the capital-productivity variable has a negative sign consistently through both model sets (with the exception of the utilities sector), and in the OLS chemicals and textiles models, it is also significant. In other words, in districts where the output–capital ratio is high, the tendency is toward little or no new investment, and this trend is particularly true of the chemicals and textiles sectors. The implication is that districts with low output–capital ratios, that is, districts with capital intensive investments are preferred.[5] This is not puzzling as it is an indirect confirmation of the cumulative causation thesis, which suggests that capital-intensity will tend to favour geographical clustering of industry.

Of the labour variables, the variable with consistent significance is district population (POP-DISTRICT), which plays a positive role in attracting new industry (for all sectors except textiles), but has little influence in determining the quantity of new investment (a small positive effect is noted for heavy industry, and, not surprisingly, a negative effect is seen for utilities). The size of the available industrial labour force has some significance, but only in the logistic models. Labour productivity (PROD-LABOUR) is inconsistently significant—primarily for all industry and the chemicals sector, and to a lesser extent the utilities sector—where it has a positive role in both attracting new investment and on the quantity of new investment. Following the discussion in the preceding paragraph, this implies that capital-intensity of existing industry often plays a key role in attracting new investment.

The infrastructure variables have little overall significance in determining the location or quantity of new industrial investment. Physical infrastructure (which is an index of access to national highways, ports, and airports), appears to have some significance in the all industry, textiles, and utilities logistic models, but none in any of the OLS/Heckman selection models. This is a somewhat unexpected finding. Perhaps this finding is an artifact of the way the INFRASTRUCTURE index has been constructed. That, however, is unlikely, as this index was chosen after an extensive sensitivity analysis among a large number of indexes, including the option of using the three infrastructure elements independently. This finding is consequential because, in policy circles, physical infrastructure is considered the key to attracting new investments (see the much discussed India Infrastructure Report, Expert Group on the Commercialization of the Infrastructure Projects, 1996). Yet, the models indicate that infrastructure does not play a role that is independent of concentration and clustering processes. In other words, the establishment of new physical infrastructure is by itself unlikely to generate new industrial investments, especially in lagging regions.[6]

The social infrastructure variables (literacy and infant mortality), not surprisingly, play no role in the new investment location decision. Literacy, whether significant or not, is negatively related, and infant mortality is unrelated. One can understand, even if one cannot endorse, the unexpected literacy effect. Statistically, it is probably the effect of Kerala's combination of high literacy and low investment. But why should infant mortality have no effect? After all there is great variation in the levels of infant mortality, which averages around 60 in the south, rising to 100 in Rajasthan, 120 in Uttar Pradesh, and 150 plus in Madhya Pradesh and Orissa. One can hope that this is a statistical artifact and not an indication of a social reality where new investments seek locations with high infant mortality. At the same time, the socialist variable (a proxy for unionized labour, or the militancy of unionized labour) is negatively related to both the existence and quantity of new investment. The implications of this are discussed in the next section.

Of the two spatial dummy variables, the METROPOLITAN dummy has no significance in any of the models, whereas the COASTAL dummy is weakly significant in some of the logistic models, and strongly significant in the OLS/Heckman selection models for all industry, chemicals, and utilities. The coastal bias of new investments (despite the absence of investments in Kerala) has already been noted in the maps and tables; similarly, the finding on metropolises is unsurprising given that we have already noted the relative decline of metropolitan districts.

THE PRIVATE AND PUBLIC SECTORS

SUMMARY INFORMATION ON THE NEW INVESTMENTS

The new or post-reform investments, as identified from the CMIE data, total just over 7 trillion Indian rupees (not including the direct investments made by state/local governments, which have been ignored throughout this analysis). Of this, 50 per cent is by the domestic private sector, 7.3 per cent is Foreign Direct Investment (FDI), 30.7 per cent is by the central government, and 12 per cent is in the joint sector which comprises private–public partnerships (see Figure 2.6 for details). For the purposes of the analysis here, the domestic private sector and FDI are added together to comprise the private sector. At only 7 per cent, FDI alone is a small fraction of the total investment pie, and is not suitable for meaningful analysis (we have conducted the analysis, and the results can be made available on request). The joint sector data, which belongs in neither of our exclusive categories, have also been omitted from the analysis. Hence the analysis that follows uses the private sector with over 57 per cent and the central government with about 31 per cent of the new investment.

Table 3.3 and Figures 2.9 and 2.10 in Chapter 2 provide some clues on the spatial distribution of these investments. The figures indicate that the geographical pattern of the private and central government investments are quite different—private investments appear to favour the coasts, and the western and southern states, and are virtually absent in Bihar, eastern Uttar Pradesh, and Assam; central government investments, on the other hand, appear to favour the eastern region, especially the inland districts of Bihar and Madhya Pradesh. The data in Table 3.3 confirm both the wider spatial coverage of the private sector (seen in Figure 2.9), and the very strong coastal bias (almost half the total private investments are in the coastal districts). The metropolitan data are unclear; certainly the intensity of investments in these districts is far higher than the non-

Table 3.3: Summary investment statistics by location

	Private Sector	Central Government
Number of districts with investment	294	164
Average investment in receiving districts	13.55	11.40
All India per-district investment	9.84	4.61
Metropolitan Districts		
Number of districts with investment	17	14
Average investment per receiving district	40.14	25.14
Share of total sectoral investment (%)	17.13	18.82
Non-Metropolitan Districts		
Number of districts with investment	277	150
Average investment per receiving district	11.92	10.12
Share of total sectoral investment (%)	82.87	81.18
Coastal Districts		
Number of districts with investment	48	32
Average investment per receiving district	40.82	22.00
Share of total sectoral investment (%)	49.18	37.65
Inland Districts		
Number of districts with investment	246	132
Average investment per receiving district	8.23	8.84
Share of total sectoral investment (%)	50.82	62.35

Source: Authors' calculations.
Note: The data sources are discussed in the text. The investment averages are in billion Rupees (in November 2006, 1 US dollar = 45 IN rupees).

metropolitan averages for both private and central sectors, but the less than complete coverage of eligible districts appears to indicate some dispersal away from these districts.

THE MODELS AND FINDINGS

The models in this section follow the same basic format of the previous analysis but with more detail and some variation. Keeping in mind that spatial analysis is a primary focus of this book (that is one of the main goals is to examine the role of spatial factors in directing new investment), it is possible to separate the purely spatial effects from the structural effects. Therefore equation 1 used in a previous section may be considered the 'full' or 'combined' model, which can be further decomposed into a 'spatial' model (equation 2) and a 'structural' model (equation 3), as given below. We did not estimate the 'spatial' and 'structural' models separately in the first part of the analysis. In this section, however, we will use these two separate models, in addition to the combined model.

$$I_{new} = C + S + e \qquad (2)$$
$$I_{new} = C + K + L + I + R + e \qquad (3)$$

Table 3.4: Determinants of probability of receiving private sector investment

Variable	Structural Model	Spatial Model	Full Model
OLDINV	0.238 ***		0.170 ***
	(23.46)		(11.01)
NEWINV-LAG		0.574 ***	0.429 ***
		(56.73)	(22.98)
IND-CREDIT	1*10⁻⁴		4*10⁻⁴
	(0.13)		(0.46)
PROD-CAPITAL	−0.363		−0.405
	(1.61)		(1.89)
POP-DISTRICT			0.663 ***
			(7.70)
LABOUR-INDUSTRY	0.136 ***		0.093 *
	(7.11)		(2.77)
PROD-LABOUR	0.002 **		0.003 **
	(4.37)		(5.86)
INFRASTRUCTURE	0.248 *		0.353 **
	(2.73)		(4.52)
LITERACY	−0.013		−0.003
	(1.15)		(0.06)
INFANT-MORT	−0.002		−0.002
	(0.22)		(0.16)
SOCIALIST	−1.120 **		−1.189 *
	(4.64)		(3.67)
COASTAL		−0.097	−0.733
		(0.05)	(1.57)
METROPOLITAN		5.889	4.109
		(0.17)	(0.11)
Constant	−2.251 **	−1.069 ***	−7.957 ***
	(5.24)	(16.69)	(13.68)
Chi-square	125.42	85.61	172.32
Correctly predicted receiving districts	92.9%	92.5%	91.5%

Source: Authors' calculations.
Notes: Number of districts with non-zero private sector investment = 292
Number of districts with zero private sector investment = 113
Figures in parentheses are Wald statistics.
*** Significant at 1 per cent
** Significant at 5 per cent
* Significant at 10 per cent

Table 3.5: Determinants of probability of receiving central government investment

Variable	Structural Model	Spatial Model	Full Model
OLDINV	0.317 ***		0.308 ***
	(19.31)		(17.99)
NEWINV-LAG		0.192 **	0.071
		(4.86)	(0.50)
IND-CREDIT	$2*10^{-4}$		$2*10^{-4}$
	(0.75)		(0.72)
PROD-CAPITAL	0.356		0.342
	(1.67)		(1.54)
POP-DISTRICT		0.570 *	−0.055
		(3.50)	(0.02)
LABOUR-INDUSTRY	0.063 **		0.063 *
	(3.94)		(3.74)
PROD-LABOUR	$2*10^{-4}$		$3*10^{-4}$
	(0.19)		(0.20)
INFRASTRUCTURE	0.022		0.028
	(0.03)		(0.05)
LITERACY	−0.007		−0.006
	(0.41)		(0.32)
INFANT-MORT	−0.002		−0.001
	(0.19)		(0.17)
SOCIALIST	0.843 *		0.751
	(3.07)		(2.25)
COASTAL		1.544 **	−0.025
		(5.35)	(0.00)
METROPOLITAN		0.192 **	0.071
		(4.86)	(0.50)
Constant	−5.084 ***	−0.934 ***	−5.131 ***
	(17.18)	(19.96)	(17.39)
Chi-square	99.62	23.39	100.13

Source: Authors' calculations.
Notes: Number of districts with non-zero central government investment = 164
Number of districts with zero central government investment = 241
Figures in parentheses are Wald statistics.
*** Significant at 1 per cent
** Significant at 5 per cent
* Significant at 10 per cent

Table 3.6: Determinants of quantity of private sector investment (N = 292)

Variable	Structural Model	Spatial Model	Full Model
OLDINV	0.181 ***		0.161 ***
	(3.60)		(3.33)
NEWINV-LAG		0.310 ***	0.225 ***
		(4.73)	(3.51)
IND-CREDIT	5*10⁻⁵		2*10⁻⁶
	(0.44)		(0.02)
PROD-CAPITAL	−0.034		0.059
	(0.10)		(0.19)
POP-DISTRICT	0.375 **		1.453
	(2.05)		(0.81)
LABOUR-INDUSTRY	0.079 ***		0.063 ***
	(3.25)		(2.63)
PROD-LABOUR	0.001 ***		0.001 **
	(2.62)		(2.35)
INFRASTRUCTURE	0.228 **		0.162
	(2.13)		(1.52)
LITERACY	0.005		−0.0005
	(0.56)		(0.06)
INFANT-MORT	0.006 *		0.006 **
	(1.96)		(2.35)
SOCIALIST	−0.914 **		−0.826 *
	(2.08)		(1.93)
COASTAL		1.236 ***	0.965 ***
		(4.29)	(3.18)
METROPOLITAN		1.162 **	0.609
		(2.54)	(1.23)
Constant	−1.568	4.11 ***	−0.276
	(1.07)	(14.18)	(0.19)
F (significance)	10.22 (.00)	24.77 (.00)	11.13 (.00)
R-square (adjusted)	0.239	0.196	0.310

Source: Authors' calculations.
Notes: This is an OLS regression model. The dependent variable is LOG of private sector investment (including FDI).
*** Significant at 1 per cent (two-tailed)
** Significant at 5 per cent (two-tailed)
* Significant at 10 per cent (two-tailed)

Table 3.7: Determinants of quantity of central government investment (N = 164)

Variable	Structural Model	Spatial Model	Full Model
OLDINV	0.108		0.077
	(1.22)		(0.89)
NEWINV-LAG		0.273	0.337 ***
		(2.25)	(2.63)
IND-CREDIT	$3*10^{-4}$ **		$3*10^{-4}$ *
	(2.29)		(1.91)
PROD-CAPITAL	0.072		0.109
	(0.20)		(0.32)
POP-DISTRICT	−0.167		−0.335
	(0.55)		(1.11)
LABOUR-INDUSTRY	−0.023		−0.042
	(0.68)		(1.21)
PROD-LABOUR	0.001 **		0.001 *
	(2.05)		(1.70)
INFRASTRUCTURE	−0.093		−0.113
	(0.54)		(0.67)
LITERACY	−0.005		−0.007
	(0.35)		(0.45)
INFANT-MORT	$-1*10^{-4}$		$9*10^{-5}$
	(0.02)		(0.02)
SOCIALIST	−0.072		−0.583
	(0.13)		(1.04)
COASTAL		0.332	0.540
		(0.83)	(1.21)
METROPOLITAN		0.708	0.532
		(1.20)	(0.77)
Constant	4.084 ***	4.750	6.408 **
	(2.68)	(15.14)	(2.48)
F (significance)	1.73 (.08)	4.13 (.00)	2.30 (.00)
R-square (adjusted)	0.043	0.055	0.094

Source: Authors' calculations.
Notes: This is an OLS regression model. The dependent variable is LOG of central government investment.
*** Significant at 1 per cent (two-tailed)
** Significant at 5 per cent (two-tailed)
* Significant at 10 per cent (two-tailed)

The Logistic Models

In all three types of models (that is, 'structural', 'spatial', and 'full'), those for the private sector have far greater explanatory power than those for the central government. The Chi-square values are much higher, as are the percentages of correctly predicted non-zero new investment districts. The spatial model for central government investments is especially weak (though all the spatial explanatory variables are statistically significant), with the lowest Chi-square value and by far the lowest percentage of correctly predicted non-zero central government investment receiving districts. The strength of the private sector models relative to the central government models indicates that the private sector is more strongly guided by the principles of investment location. Investment in infrastructure items like power, which is what the central government specializes in, tends to be either natural resource-oriented (seeking mines or running water) or politically motivated.

The two most important determinants of success or failure for private investment, that is, whether or not a district receives any new private sector investment are: the existence of any investment in the pre-reform era (OLDINV), and the existence of new private investment in the neighbouring districts in the post-reform era (Spatial Lag). Both variables are highly significant in the separate 'structural' and 'spatial' models, and in the 'combined' model. On the other hand, the Spatial Lag term is not significant for central government investments (full model), implying that clustering effects are weaker in this case. Similarly, the OLDINV variable has the expected but weak effect in the central government models.

The set of Labour variables (population size, size of manufacturing labour force, and labour productivity) are all significant for the private sector models, indicating that labour considerations play a significant role in the private sector location decision. In the central government models, labour is a much weaker consideration—the district population size is significant, as is, to a considerably lesser extent, the size of the manufacturing labour force, but labour productivity is of no consequence in the central government models.

The role of infrastructure is as expected. The literacy and infant mortality levels have little bearing on whether a district receives private sector or central government investment. The availability of physical infrastructure, on the other hand, plays a weak positive role in attracting private sector investment, but has no bearing on locating central government investment. Finally, private investment tends to avoid socialist states, but central government investments appear to be indifferent to local political orientation.

The OLS Models

The OLS regression models for the private sector are strong and robust; for the central government they are weak, with little explanatory power. The adjusted-R-square values range from 0.20 to 0.31 for the private sector models. In the central government models, these numbers range from 0.04 to 0.09, suggesting that the quantity of new investment tends to be guided far more strongly by market factors for the private sector than for the central government.

The two most revealing trends of the logistic models are further confirmed here: First, the two most significant predictors of the quantity of new private investment are OLDINV and SPATIAL-LAG, whereas in the central government models, OLDINV is not significant. In other words, the quantity of existing investment in district i and the quantity of new private investment in the neighbours of district i, are the most important predictors of the quantity of new private sector investment in district i. Second, labour characteristics are significant in predicting the quantity of new private sector investments, but not for central government investments. In fact, population size and manufacturing labour force size have a counter-intuitive sign (though not statistically significant) in the central government models.

The infrastructure variables generally have the least explanatory power in both sets of models. In the central government models, none of the infrastructure variables are significant in either the 'structural' or 'full' variation. Physical infrastructure is significant in the private sector 'structural' model, but not in the 'full' model (the Coastal variable is probably picking up much of the explanatory power). The unexpected relationship with infant mortality seen in the previous section is repeated for the private sector models here.

There may be an interesting story on labour emerging here. It appears that aside from the spatial factors discussed above, a primary

DETERMINANTS OF INDUSTRIAL LOCATION

consideration of private capital relates to the issue of labour. First consider what is not important. The two social infrastructure variables (literacy and infant mortality) have little or no effect on directing new investment—the parameter estimates are generally not significant, and often have a counter-intuitive sign. The Physical Infrastructure variable is significant but weak. On the other hand, the Labour variables (size of industrial labour, population size, and labour productivity) generally show positive and significant effects in the private sector models. At the same time, the Socialist variable (a proxy for unionized labour, or the militancy of unionized labour) [7] is negatively related to new investment.[8]

Perhaps the models are hinting at some story about the kind of labour that private capital seeks. Industrial labour presence is favored, in quantity and quality, but not if it is strongly unionized; the labour need not be literate, and given that high infant mortality is indicative of poor social conditions and social infrastructure, the social conditions of labour need not be progressive. The evidence presented here is circumstantial—indeed, these models were not designed to examine the relationship between labour conditions and new investment—but they raise significant questions for further research

The private sector 'spatial' model is remarkably effective. Note that the right hand side of this model has a single continuous variable (SPATIAL-LAG) and two dummy variables. This simple model is able to explain about 20 per cent of the variation in the distribution of new investment quantity. All the spatial factors are significant, and compared to its counterpart 'structural' and 'full' models, the constant term is meaningful, and the F-statistic is much higher. It is a simple model that highlights the importance of the coastal and metropolitan factors, and compared to the far less effective central government model, highlights the importance of geographical factors for the private sector.

WHAT REALLY MATTERS IN INDUSTRIAL LOCATION

The story that emerges from the analysis in this chapter is clear in some parts and murky in others. It is clear that both in determining which districts get some new investment and in determining the quantity of new investment, the most significant factors are the existence and size of investment from the pre-reform period and the existence and size of new investment in the neighbouring districts. The first factor implies continuity—evidence of a historical process of investment location. The second factor implies clustering—evidence of the role of geography in guiding investment location. Both factors are stronger in determining success/failure rather than quantity. One implication of this is that though historical processes are being continued in the choice of investment location, the volume of new investments is following a somewhat different pattern from the past. In other words, districts that were successful earlier continue to receive new investment, but degree of past success is not the best indicator of the degree of current success. The most successful pre-reform districts are not the most successful post-reform districts. There has been a shift in geographical focus whereby new investments seek locations within the existing leading regions (or clusters), but at new locations within these regions. To use a concrete example highlighted in Chapter 2: Greater Bombay is still successful in attracting investment, but not to the extent it was earlier; its neighbours, Raigarh and Thane, are now the preferred investment destinations.

It is also clear that we are unable to identify all the factors that influence industrial location decisions. There are non-random local factors that have not been modelled here, and that may indeed be difficult to model. These factors relate to local or state level policies such as tax incentives, the location of export processing and/or free trade zones, etc. Consider the most obvious of these: the district level incentive system that is common in most states. States have a three to five tier incentive system whereby there are tax breaks for locating in lagging districts (typically lagging districts are also differentiated by the degree to which they are behind the leading districts). For instance, West Bengal and Maharashtra have three- and five-tier systems in which the metropolitan area is typically exempted from incentives. It is difficult to use this information in a modelling framework, mainly because the incentive structures vary so much from state to state. This is a serious policy issue. We will take it up again in Chapter 6.

Moreover, there is surely a substantial random element in the distribution. Recall the survey-based finding that personal preference or chance is the most common factor in the location decision. In addition, there are intangible elements like culture, entrepreneurship, and initiative. For instance, one of the reasons often cited (in non-academic circles) for Gujarat's recent success is the entrepreneurial culture of the local population. This was supposedly kept under check (though not completely) in the previous regime of licenses and controls, but with the removal of these checks, the state's entrepreneurial potential is being better realized. This kind of thesis is probably best tested using other methodologies, for it is difficult to create an objective index of entrepreneurial culture. Nevertheless, it is important to understand that the location decision of a firm does not rest entirely on objective, measurable conditions.

The bottom line is that there is increasing divergence between states and districts on industrialization. We have identified the principal source of the recent divergence—which is the increasing dominance of private capital. We have argued that private sector investment location decisions are based on profit-maximizing or efficiency related factors such as access to markets, capital, labour, infrastructure, and so on, whereas the central government investment location decisions are less influenced by such efficiency related factors. Furthermore, we have suggested that in seeking efficiency advantages, private sector investments will tend to locate in existing industrial clusters and metropolitan centres, with access to the coast, and avoid regions with inhospitable local governments.

The evidence presented here provides unambiguous support for both propositions. Private sector investment location decisions are indeed guided by efficiency related factors to a far greater extent than are location decisions by the central government. Also, private sector investments are seen to favour existing industrial clusters (providing support for the idea that the already leading industrial regions would benefit most), and coastal districts, and are seen to be averse to communist or socialist states. There is less support for the argument that such investments will also favour metropolitan regions.[9] On the other hand, central government investments appear not to be guided by any clear spatial considerations. These findings are consistent in both the modelling frameworks—success/failure, and the quantity of new investment, that is, in determining whether or not a district gets new investment, and in determining the quantity of new investment. In other words, it is possible to discern some of the locational logic of private capital: It favours existing industrial areas and tends to form clusters. There is far less evidence that central government location decisions follow the same locational logic. These two logical systems of investment distribution have, in combination, created the conditions of increasing industrial divergence in India.

In this chapter, we have been able to identify the core factors that influence investment distribution at the district level and can therefore argue that increasing privatization of industry will continue to lead to increasing industrial divergence between states and districts, at least in the short run. In the next chapter we look at investment location decisions at the firm level and identify the factors of economic geography that influence industrial location at the micro scale.

NOTES

1. Markusen *et al.* (1991) have discussed the regional impacts of US defence expenditures. Naughton (1988) provides evidence from China and its pre-1978 'third front' strategy that privileged the interior regions not only for the defence industry but for the manufacturing industry in general.
2. Most studies have been limited to estimating the effect of state regulations (on the environment for instance) or incentives (tax breaks, accelerated depreciation rates, etc.) on the location behaviour of firms.
3. The data definitions and sources, unless mentioned otherwise, are as follows. Literacy: from the 1991 Population Census, reported in 'Profiles of Districts', defined as the percentage of the population that is literate. Infant Mortality: from Rajan and Mohanchandran (1998), defined as the number of deaths per 1000 live births at age 5 years, estimated from the 1991 Population Census. Manufacturing Labour: from the 1991 Population Census, reported in 'Profiles of Districts', defined as the percentage of workers employed in non-household manufacturing industries. Industrial Credit: reported in 'Profiles of Districts', defined as the per capita bank credit to industries derived from the information on scheduled commercial bank branches, deposits and credits on the last Friday of March 1993. 'Profiles of Districts' is a 1993 CMIE publication.
4. Here spatial lag has been calculated as the log of the sum of the neighbouring I_{new} rather than from the raw investment quantity (to mitigate the problem of lack of normality in the raw data).
5. Low output–capital ratios do not necessarily imply capital intensity. Plants using new technologies may, for a time, be capital intensive and have high output–capital ratios. Hence, the link between capital intensity and geographical clustering we make here is only suggestive, a by-product, if you will, of the analysis.

6. Does the inclusion of Physical Infrastructure as an explanatory variable introduce a problem of endogeneity? In other words, are both Physical Infrastructure and new investment endogenously determined in a larger system (represented by the rest of the explanatory factors identified here), and is there a reciprocal causal connection between the two variables? Had the two data series been taken from the same time period, that is, had Physical Infrastructure been represented by new investments in roads, ports, etc., the problem of endogeneity would have been a serious one. However, as used here, the Physical Infrastructure data are historic, indexing a cumulation of infrastructure access indicators that predate the new investment data. There may indeed be a circular and cumulative relationship between the two variables (in fact, this relationship is theorized at the outset), but as analysed here there is no simultaneous relationship, nor are the determining factors the same.
7. West Bengal accounted for about 38 per cent and Kerala for over 5 per cent of the workdays lost nation-wide in dispute related stoppages between 1985 and 1995 (calculated from Government of India, 1997).
8. The behaviour of the Socialist variable is similar for the private sector in both logistic and linear regressions (a significant negative relationship), and insignificant for the central government (though the sign does change).
9. As pointed out in Chapter 2, this either provides support for the thesis that liberalization will break the monopoly power of metropolises or this indicates that environmental regulations that prohibit the location of polluting industry in metropolitan areas are effective.

4

Economic Geography and the Firm

In the previous chapter, we identified the location-based factors that attract industrial capital to specific districts. In this chapter, we turn to the firm and consider the question of how and to what extent location based features have cost effects at the firm level. That is, are there cost benefits to firms that arise purely as a result of the spatial arrangements of other firms (in the same industry and in other industries) and transportation infrastructure. In Chapter 1, we had pointed out that the 'New Economic Geography' literature (well reviewed in Fujita, Krugman, and Venables 1999) allows us to move from the question 'Where will manufacturing concentrate (if it does)?' to the question 'What manufacturing will concentrate where?' Now we wish to go beyond these questions, to ask, 'What manufacturing will locate where and *why*'?

To understand the process of industrial location and concentration, it is important to first analyse the location decisions of firms in particular industries. The location decision of the individual firm may be influenced by several factors. These include:

(1) availability of infrastructure, access to markets, and the external economies provided by localization and urbanization, that is the 'economic geography';
(2) local wages, taxes, subsidies, and incentives, and the general policy environment, that is the 'political economy';
(3) history, being 'accidental';

Here, we focus on the economic geography characteristics. We develop and estimate an economic model to assess the impact of region specific characteristics on location choices of firms in well-defined industries. For the empirical application, we use micro-level establishment data for Indian industry to examine the contribution of regional characteristics on location choices. Our concept of regional characteristics extends beyond its natural geography. Rather than focusing on inherent characteristics such as climate and physical distance to the coast and market areas, we analyse the economic geography of the region.

Recall from the discussions in Chapter 1 that economic geography characteristics include two elements: market access, represented by the transport network linking a location to market centres; and spatial externalities, represented by the local presence of buyers and suppliers to facilitate inter-industry transfers, the local presence of firms in the same industry to facilitate intra-industry transfers, and the diversity of the local industrial base. Drawing on testable hypotheses from the NEG literature, this analysis provides the micro-foundation for understanding whether a region's economic geography influences location decisions at the firm level. Only by first explaining these decisions, will it be possible to build a general framework for evaluating the overall spatial distribution of economic activity and employment.

Using plant or 'factory' level data for 1998-9, from the Indian Annual Survey of Industries (ASI), we examine location choices in eight manufacturing industries.[1] These are (with National Industrial Classification [NIC] codes in parenthesis):

(1) Food processing (151, 152, 153, 154, 155)
(2) Textiles and textile products, including wearing apparel (171, 172, 173, 181)
(3) Leather and leather products (191, 192)
(4) Paper products, printing and publishing (210, 221, 222)
(5) Chemical, chemical products, rubber and plastic products (241, 242, 243, 251, 252)
(6) Basic metals and metal products (271, 272, 273, 281, 289)
(7) Mechanical machinery and equipment (291, 292)
(8) Electrical and electronics (including computer) equipment (292, 300, 31, 32)

These plant level data are supplemented by district and urban demographic and amenities data from the 1991 Census of India, and detailed, geographically referenced information on the availability and quality of transport infrastructure linking urban areas (CMIE 1998;

ML Infomap 1998). The ASI data allow us to identify each plant at the district level spatially and at the three digit SIC level sectorally.

This chapter is organized into five sections with an appendix. In the first section, we present the analytic framework. The econometric model to examine location decisions at the firm level is specified in an appendix to this chapter. It is a technical discussion that does not need to be placed in the narrative of this chapter. In the second section, we identify the economic geography variables that are expected to have an influence on a firm's cost structure. In the third section, we discuss some data issues and present summary statistics on spatial concentration. In the fourth we discuss results from the econometric analysis, and we briefly summarize the contributions and implications of the findings in the final section.

THE ANALYTIC FRAMEWORK

The analytic framework to examine the location of manufacturing industry primarily draws on recent findings from the NEG literature. In the NEG literature, Krugman (1991a, 1991b) and Fujita et al. (1999) analytically model increasing returns, which stem mostly from pecuniary externalities.[2] They emphasize the importance of supplier and demand linkages and transportation costs. Firms prefer to produce each product in a single location given fixed production costs, and firms also prefer to locate their production facilities near large markets to minimize transportation costs.

Drawing upon Fujita and Thisse (1996) and Fujita (1988), we model firms to benefit from externalities arising from being co-located with other firms. If $a(x, y)$ is the benefit to a firm at x obtained from a firm at y, and $f(y)$ denotes the density of firms at each location $y \in X$ then,

$$A(x) \equiv \int_x a(x, y) f(y) dy \qquad (1)$$

Thus, $A(x)$ represents the aggregate benefit accrued to a firm at x from the externalities created in location X. Assuming that production utilizes land (S_f) and labour (L_f) with rents and wages of $R(x)$ and $W(x)$ respectively at x, a firm located at $x \in X$ would maximize profits subject to:

$$\Pi(x) = A(x) - R(x)S_f - W(x)L_f \qquad (2)$$

Note that as an aggregate term, the density of firms at each location, $f(y)$, can represent regional economic attributes based on inter-firm relationships (in other words, economic geography). Specifying types of such attributes unpacks the sources of spatial externalities, which have been often treated as a black box in neoclassical urban system models (Henderson 1988). First, a large geographic concentration of similar firms can provide scale economies in the production of shared inputs. Besides, firms that utilize similar technologies and face common issues are more likely to collaborate with one another to share information on a variety of issues from problem solving to the development of new production technologies. Second, the benefits from locating near own-industry concentrations can be augmented by the presence of interrelated industries.

To a large extent, the work on inter-industry externalities have been motivated by research on industry clusters. Clusters can be defined as a geographically concentrated and interdependent network of firms linked through buyer–supplier chains or shared factors. The success of an industry cluster hinges on how well such local linkages among firms, education and research institutions, and business associations can be developed. The 'cluster' concept particularly emphasizes interfirm relations that reduce the cost of production by lowering transaction costs among firms (Porter 1990). Interrelated firms located in proximity can reduce their transportation cost for intermediate goods and can share valuable information on their products more easily. Therefore, for profit maximizing firms, the presence of a well-developed network of suppliers in a region is an important factor for their location decision. Lastly, economic diversity of a region is another important source of spatial or location-based externalities. Firms located in larger metro areas are more likely to benefit not only from inter-industry technology spillovers but also from easier access to producer services such as legal services or banking.

Transport costs are also important in determining the location choice of firms. Krugman (1991b) shows that manufacturing firms tend to locate in regions with larger market demand to realize scale economies and minimize transportation cost. If transport costs are very high, then activity is dispersed. In the extreme case, under autarky, every location must have its own industry to meet final demand. On the other hand, if transport costs are negligible, firms may be randomly distributed as proximity to markets or suppliers will not matter. Agglomeration would occur at intermediate transport costs when the spatial mobility of labour is low (Fujita and Thisse 1996). We therefore expect a bell shaped (inverted U-shaped) relationship between the extent of spatial concentration and transport costs (see Chapter 1 and Figure 1.1).

To include transport costs in a firm's location decision, we modify equation (2) as:

$$\Pi(x) = A(x) - R(x)S_f - W(x)L_f - TC(x) \qquad (3)$$

where $TC(x)$ represents the transport costs of the firm at location x. With a decline in transport costs, firms have an incentive to concentrate production in a few locations to reduce fixed costs. Transport costs can be reduced by locating in areas with good access to input and output markets. Thus, access to markets is a strong driver of agglomeration towards locations where transport costs are such that it is relatively cheap to supply markets. In addition to the pure benefits on minimizing transport costs, the availability of high quality infrastructure linking firms to urban market centres increases the probability of technology diffusion through interaction and knowledge spillovers among firms, and also increases the potential for input diversity (Lall *et al.* 2004a). Analytical models of monopolistic competition generally show that activities with increasing returns at the plant level are pulled disproportionately towards locations with good market access.

The analytic framework in this section highlights the importance of economic geography in influencing location and agglomeration at the firm level. Insights from NEG and regional science models suggest that own- and inter-related- industry concentrations, availability of reliable infrastructure to reduce transport costs and enhance market access, regional amenities and economic diversity are important for reducing costs, thereby influencing location and agglomeration of industry. In the following section, we describe the economic geography variables that are used in this analysis. The econometric specification to evaluate the importance of these variables is described in an appendix to this chapter. The empirical strategy is to estimate a cost function to see how costs (and thereby profits) are affected by the economic geography of the region where the firm is located. If specific factors related to the local economic geography have cost reducing impacts, then firms are likely to choose regions with disproportionately higher levels of these factors.

ECONOMIC GEOGRAPHY VARIABLES

OWN-INDUSTRY CONCENTRATION

The co-location of firms in the same industry (localization economies) generates externalities that enhance productivity of all firms in that industry. These benefits include sharing of sector specific inputs, skilled labour, and knowledge, intra-industry linkages, and opportunities for efficient subcontracting. Firms that share specialized inputs and production technologies are more likely to cooperate in a variety of ways. In many industries, it is common for competitors in the market to launch joint projects for new product and process development. Further, a disproportionately high concentration of firms within the same industry increases possibilities for collective action; for example, to lobby regulators or influence bid-prices of intermediate products See the discussion in Chapter 1 and the significance of the theoretical works of Marshall (1890), Arrow (1962), and Romer (1986) and the empirical literature supporting the positive effects of localization economies (Henderson 1988; Ciccone and Hall 1996).

There are several ways of measuring localization economies. These include own-industry employment in the region, own-industry establishments in the region, or an index of concentration, which reflects disproportionately high concentration of the industry in the region in comparison to the nation. We use own-industry employment in the district to measure localization economies. This measure is consistent with the type of benefit spillovers specified in equation (1), where localization economies come from the absolute volume of other activity in the district. Own-industry employment is calculated from employment statistics provided in the 1998–9 sampling frame of the ASI, which provides employment data on the universe of registered industrial establishments in India. The sample data used for the cost function estimation are drawn from this sampling frame.

Inter-industry Linkages

In addition to intra-industry externality effects, we also include a measure to evaluate the importance of inter-industry linkages in explaining firm level profitability, and thereby location decisions. The importance of inter-industry linkages in generating localization forces was first introduced by Marshall (1890, 1919). Recent work by Venables (1996) has shown that agglomeration could occur through the combination of firm location decisions and buyer–supplier linkages even without high factor mobility. The presence of local supplier linkages makes buyer industries more efficient and reinforces the localization process.

There are several approaches for defining inter-industry linkages: input–output based, labour skill based, and technology flow based. Although these approaches represent different aspects of industry linkages and the structure of a regional economy, the most common approach is to use the national level input–output accounts as templates for identifying strengths and weaknesses in regional buyer–supplier linkages (Feser and Bergman 2000). The strong presence or lack of nationally identified buyer–supplier linkages at the local level can be a good indicator of the probability that a firm is located in that region.

To evaluate the strength of buyer–supplier linkages for each industry, we use the summation of regional industry employment weighted by the industry's input–output (IO) coefficient column vector from the national input–output account:

$$L_{ir} = \sum_{i=1}^{n} \omega_i e_{ir} \tag{4}$$

where L_{ir} is the strength of the buyer–supplier linkage, ω_i is industry i's national input–output coefficient column vector and e_{ir} is total employment for industry i in district r. This is similar to the measure used by Koo (2002) to define the strength of buyer–supplier chains. The measure examines local level inter-industry linkages based on the national input-output accounts. The national input–output coefficient column vectors describe intermediate goods requirements for each industry (that is, inter-industry linkages). Assuming that local industries follow the national average in terms of their purchasing patterns of intermediate goods, national level linkages can be transposed to the local level industry structure for examining whether region r has a right mix of supplier industries for industry i. By multiplying the national input–output coefficient column vector for industry i and the employment size of each sector in region r (a district is used as a geographical unit for buyer–supplier linkage analysis), simple local employment numbers can be weighted, based on what industry i purchases nationally.

Indeed, the importance of local linkages is determined by the size of its industrial base (for example, employment in each industry) and the extent to which local industries can provide intermediate goods for local firms (from the input–output coefficient vector). In this case, our measure takes two important aspects of buyer–supplier linkages into account—fit and size. While computing the indicator, we noticed that the industry categories in the NIC system and in input–output accounts do not have an exact match. Therefore, we first developed a concordance table between them before multiplying w_i and e_{ir}. Data on input-output transactions are from the Input–Output Transactions Table 1993–94, Ministry of Statistics and Programme Implementation.

Economic Diversity

In addition to buyer–supplier linkages, there are other sources of inter-industry externalities. Prominent among these is the classic Chinitz–Jacobs diversity. The diversity measure provides a summary measure of urbanization economies, which accrue across industry sectors and provide benefits to all firms in the agglomeration. Chinitz (1961) and Jacobs (1969) proposed that important knowledge transfers primarily occur across industries and the diversity of local industry mix is important for these externality benefits. Therefore, industries with Jacobs-type externalities tend to cluster in more diverse and larger metro areas. Recently, Duranton and Puga (1999) designed a model providing the micro-foundations of a Jacobs-type model. Also see the discussion in Chapter 1, where we show that the benefits of locating in a large diverse area go well beyond the technology spillovers argument, arising from better access to business services, the heterogeneity of economic activity, increased range of local goods available, increased output variety in the local economy, etc. The latter type of benefit is particularly important in developing countries, where most manufacturing industries are based on low skills and low wages but abundant local labour forces.

Here we use the well-known Herfindahl measure to examine the degree of economic diversity in each district. The Herfindahl index of a region r (H) is the sum of squares of employment shares of all industries in region r:

$$H_r = \sum_j \left(\frac{E_{jr}}{E_r}\right)^2 \tag{5}$$

Unlike measures of specialization, which focus on one industry, the diversity index considers the industry mix of the entire regional economy. The largest value for H_r is one when the entire regional economy is dominated by a single industry. Thus, a higher value signifies lower level of economic diversity. Therefore, for more intuitive interpretation of the measure, for the diversity index in our model, H_r is subtracted from unity. Therefore, $DV_r = 1-H_r$. A higher value of DV_r signifies that the regional economy is relatively more diversified.

MARKET ACCESS

Following theory, improved access to consumer markets will increase the demand for a firm's products, thereby providing the incentive to increase scale and invest in cost reducing technologies. The distance from, and the size and density of, market centres in the vicinity of the firm determine access to markets. The classic gravity model, which is commonly used in the analysis of trade between regions and countries (Evennet and Keller 2002), states that the interaction between two places is proportional to the size of the two places as measured by population, employment or some other index of social or economic activity, and inversely proportional to some measure of separation such as distance. Following Hansen (1959):

$$I_i^c = \sum_j \frac{S_j}{d_{ij}^b} \tag{6}$$

where I_i^c is the 'classical' accessibility indicator estimated for location i, S_j is a size indicator at destination j (for example, population, purchasing power or employment), d_{ij} is a measure of distance (or more generally, *friction*) between origin i and destination j, and b describes how increasing distance reduces the expected level of interaction. Empirical research suggests that simple inverse distance weighting describes a more rapid decline of interaction with increasing distance than is often observed in the real world (Weibull 1976). The most commonly used modified form is a negative exponential model such as:

$$I_i^{ne} = \sum_j S_j \cdot e^{(-d_{ij}^b/2a^2)} \tag{7}$$

where I_i^{ne} is the potential accessibility indicator for location i based on the negative exponential distance decay function, most other parameters are defined as before, and the parameter a is the distance to the point of inflection of the negative exponential function.

There are several options for developing accessibility indicators depending on the choice of distance variables used in the computation. These include: (1) indicators based on Euclidean distance; (2) indicators incorporating topography; (3) indicators incorporating the availability of transport networks; (4) indicators incorporating the quality of transport networks. In studies related to agglomeration economies and economic geography (for example, Hanson 1998), the distance measure of choice is usually the straight line distance, which has the advantage of computational simplicity. However, this assumption of a Christaller-type isotropic plane is clearly unrealistic, particularly in countries where topography and sparse transport networks of uneven quality greatly affect the effort required to move between different parts of the country.

A better alternative is to use network distance as the basis of the inverse weighting parameter and to incorporate information on the quality of different transportation links. Feasible travel speed and thus travel times will vary depending on each type of network link. A place located near a national highway will be more accessible than one on a rural, secondary road. The choice of the friction parameter of the access measure will therefore strongly influence the shape of the catchment area for a given point, that is, the area that can be reached within a given travel time. This, in turn, determines the size of potential market demand as measured by the population within the catchment area. Figure 4.1 shows the accessibility surface for

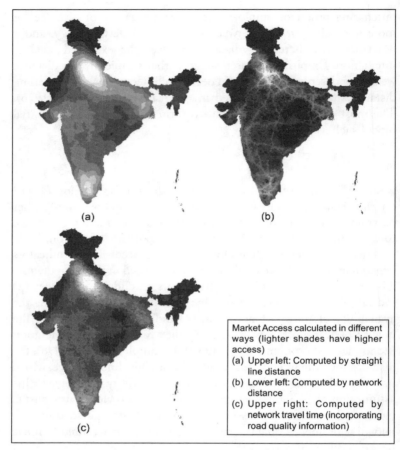

Figure 4.1: Maps of market access

Source: Our calculations from digital map of transportation infrastructure (ML Infomap 1998) and Census of India, 1991.
Note: Maps not to scale.

India using three measures of market access: (1) based on Euclidean distance, (2) network distance, and (3) network travel time. It is clear that indicators based on (1) and (2) overestimate potential market area, and the variation in infrastructure quality between regions leads to a more realistic representation of the structure of market areas. Thus, incorporating the quality of the transport network is important in assessing the potential market integration.

We computed an accessibility index, which describes market access using digital maps of the Indian road network and the location and population of urban centres (ML Infomap 1998). The urban centres database includes latitude and longitude coordinates and 1991 population for 3752 cities with a total population of about 217 million. This represents more than one quarter of India's total 1991 population of 846 million. Measures of personal income or consumption may better represent approximate market attractiveness, and employment levels may be a better indicator of the local labour pool. However, in the absence of detailed local data on these parameters, total urban population represents a reasonable proxy for potential market access.

The digital transport network data set includes an estimated 400,000 km of roads.[3] Each road segment is categorized into four classes according to road quality: National highways (about 30,000 km or 7.7 per cent of total roads), state highways (90,000/22.5 per cent), secondary connector roads (120,000/29.8 per cent), and other roads (160,000/39.9 per cent). The complete digital representation of India's transport network that is used for the accessibility index thus consists of a set of urban centres represented as nodes connected by lines that correspond to roads of different quality. Rather than distance, the weighting parameter used in the accessibility computation is an estimate of travel time. This makes it possible to incorporate road quality information by assigning different travel speed estimates to different types of roads. Based on information available in the Indian Infrastructure Handbook (CMIE 1998) varying travel speeds ranging from 25 to 50 km/hour were used, depending on the type of road. The algorithm used to compute the accessibility measure is based on the Dijkstra algorithm. We use this to compute the network travel time to urban centres for each of more than 100,000 points distributed across India. As the exact geographic location of each firm is not publicly available, we summarized the accessibility for each district by averaging the individual values for all points that fall into the district. The negative exponential function in equation (7) is chosen as the most suitable functional form for the decay of interaction with increasing travel time.

Finally, we calculated distances (travel times) to trans-shipment hubs to see if these had external effects over and beyond the effects of market accessibility. In general, trade flows through hubs are disproportionately higher than through nodes (points) along a simple linear network. As a result, proximity to hubs will provide firms with a larger choice of transport providers and intermediate input suppliers than market centres along a linear network. Further, trans-shipment nodes (such as ports) have historically had an important role in the evolution of urban centres. In fact, as a result of path dependency, such urban centres continue to be prosperous (and efficient) even after the initial advantage of the hub access becomes irrelevant (Fujita and Mori 1996). In the analysis, we use travel time to seaports as a measure of distance from hubs. Data limitations preclude us from expanding the choice of indicators to include surface transport and airport hubs. We computed travel times between each district headquarter and the closest hub using the same road network as in the market accessibility measure.

DATA AND SUMMARY STATISTICS

Data Sources

We use plant level data for 1998–9 from the ASI conducted by the CSO of the Government of India. The 'factory' or plant is the unit of observation in the survey, and data are based on returns provided by factories.[4] Data on various firm-level production parameters such as output, sales, value added, labour cost, employees, capital, materials, and energy are used in the analysis. In summary, factory level output is defined as the ex-factory value of products manufactured during the accounting year for sale. Capital is often measured by perpetual inventory techniques. However, this requires tracking the sample plant over time. This is a major task for micro-level research due to changes in sampling design and incomplete tracking of factories over time. Instead, in our study (and in the ASI dataset) capital is defined as the gross value of plant and machinery. It includes not only the book value of installed plant and machinery, but also the approximate value of rented-in plant and machinery. Doms (1992) demonstrates that defining capital as a gross stock is a reasonable approximation for capital. Labour is defined as the total number of employee person-days worked and paid for by the factory during the accounting year.

The factory or plant level data from the Indian ASI allows us to compute input costs. With respect to input costs and input prices, capital cost is defined as the sum of rent paid for land, building, plant, and machinery, repair and maintenance cost for fixed capital, and interest on capital. Labour cost is calculated as the total wage paid for employees. Energy cost is the sum of electricity (both generated and purchased), petrol, diesel, oil, and coal consumed. The value of self-generated electricity is calculated from the average price that a firm pays to purchase electricity. Material cost is the total aggregate purchase value for domestic and foreign intermediate inputs. We define the price of capital as the ratio of total rent to the net fixed capital. The price of labour is calculated by dividing total wage by the number of employees. Energy and material prices are defined as weighted expenditure per unit output. Output value is weighted by factor cost shares.

Data quality has been examined by cross referencing with standard growth accounting principles as well as by reviewing comments from

Table 4.1: Characteristics of firms in the study sectors

Location	Industry	Firms	Employment	Wages/ Employee	Output/ Employee	Value Added/ Employee
Nationwide	All industries	23201	4605	60	277	127
	Food processing	4168	671	47	253	147
	Textiles	3409	1111	44	140	76
	Leather	468	79	41	211	135
	Paper products and printing	1043	129	70	314	204
	Chemicals	2811	474	83	376	79
	Metals	2331	410	77	261	114
	Mechanical machinery	1300	237	78	189	95
	Electrical and electronics	1267	251	101	344	65
	Other industries	6404	1243	54	385	195
Non Urban		8343	1494	50	301	126
Non-Metro Urban		9446	1972	58	235	125
Metro Urban		5412	1139	74	320	133

Source: ASI 1998–9.
Note: Data for employment, wages/ employee, output/ employee and value added / employee are in thousands.

other researchers who have used these data. The geographic attributes allow us to identify each firm at the district level.

Table 4.2: Concentration in industrial sectors

Industry	NIC Code	No. of States	G	H	r
Jute Textiles	25	12	0.548	0.021	0.570
Beverages	22	23	0.313	0.019	0.329
Leather and Leather Products	29	17	0.143	0.012	0.146
Miscellaneous Food Products, n.e.c.	21	24	0.092	0.003	0.098
Wood and Wood Products	27	26	0.079	0.007	0.080
Textile Products	26	20	0.066	0.002	0.070
Wool and Silk Textiles	24	20	0.058	0.006	0.058
Food Products	*20*	*26*	*0.043*	*0.001*	*0.046*
Basic Metals and Alloys	*33*	*24*	*0.053*	*0.020*	*0.038*
Cotton Textiles	*23*	*21*	*0.029*	*0.002*	*0.030*
Chemicals and Chemical Products	*30*	*24*	*0.027*	*0.002*	*0.027*
Non-Metallic Mineral Products	32	26	0.019	0.001	0.019
Transport Equipment and Parts	37	22	0.025	0.009	0.018
Machinery other than Transport/ Electronic/Electrical	35	22	0.018	0.006	0.013
Electronic and Electrical Machinery, Parts, and Apparatus	36	24	0.018	0.009	0.010
Rubber, Petroleum and Coal Products	31	24	0.011	0.005	0.007
Metal Products	34	27	0.007	0.004	0.002
Paper and Paper Products	28	25	0.006	0.004	0.002
Mean			0.083	0.008	0.083

Source: Calculated from Annual Survey of Indian Industries sample frame, 1998–9.
Note: The most concentrated sectors are shown in bold, the group with significant but lower levels of concentration are in italics, the remainder are not concentrated.

Table 4.3: Location and productivity

Location type	No. of Firms	Labour Productivity
Rural	12378	1022.7
Non-metro Urban	24691	1163.6
Metro Urban	10255	1391.2
Total	47324	1176.0

Source: Calculated from Annual Survey of Indian Industries 1998–9.

ECONOMIC GEOGRAPHY AND THE FIRM 121

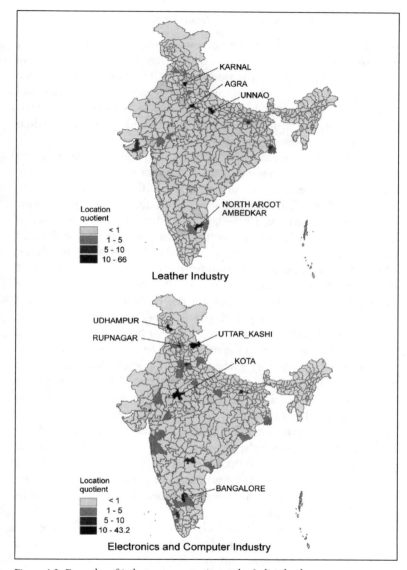

Figure 4.2: Examples of industry concentration at the 3-digit level

Source: Authors' calculations from ASI database for 1998–9.
Note: Maps not to scale.

Summary Information on Spatial Variation

Before moving on to discussing the results from the empirical analysis, we provide a general overview on the concentration and basic characteristics of firms in the study sectors. The essence of economic geography is the spatial concentration of economic activities and subsequent economic benefits. Therefore, examining spatial concentration patterns of firms is the first necessary step when investigating economic geography.

We first divide the economic landscape into non-urban areas, urban areas, and large metropolitan areas. The metropolitan areas include the following cities and their urban agglomerations—Delhi, Mumbai, Calcutta, Chennai, Bangalore, and Ahmedabad. Using the sample data from the ASI for 1998–9, we see that average wages across industries are the highest in metropolitan areas (see Table 4.1). In comparison to a nationwide average annual of Rs 60,000 per employee, labour remuneration is Rs 74,000 for metropolitan areas, Rs 54,000 for other urban areas and Rs 50,000 for non-urban areas. Among various industries, annual wages are the highest in electronics and computers (Rs 101,000 per employee) and lowest in the leather industry (Rs 41,000 per employee). Even within sectors, wages tend to be higher as we move up the urban scale (not shown in this table).

Productivity indicators such as output per employee and value added per employee show interesting patterns. While per employee output is quite high in several industries, the value added figures show quite a different situation. For example, per employee output in computing and electronics is Rs 344,000 but value added per employee is only Rs 65,000. Similarly, the numbers for output and value added per employee are Rs 376,000 and Rs 79,000 for chemicals and Rs 314,000 and Rs 204,000 for printing and publishing. This suggests that these industry sectors are not very efficient in transforming inputs into higher value outputs.

Next, we use the Ellison–Glaeser (1997) or EG index of concentration to see if industrial activity within sectors is clustered across locations. Their concentration index can be defined as:

$$r = \frac{\sum_{i=1}^{M}(s_i - x_i)^2 - (1 - \sum_{i}^{M} x_i)H}{(1 - \sum_{i}^{M} x_i)(1 - H)} \tag{8}$$

where r is the extent to which an industry is geographically concentrated, s_i is the region i's share of the study industry, x_i is the regional share of the total employment, and H is the Herfindahl industry plant size distribution index, $H = \sum_{j=1}^{N} z_j^2$. The EG index is explicitly derived from the micro-foundations of a firm's location choice. It takes on a value close to zero when the distribution of plant location is completely random (as opposed to a uniform distribution). Therefore, a non-zero value implies agglomeration or clustering above and beyond what we would observe if the firm's location decisions are random (in general, an industry is highly concentrated if r ≥ 0.05, moderately concentrated if r is between 0.02 and 0.05, and not concentrated if r <0.02). The index is designed to allow comparisons across industries, over space and over time. Therefore, in principle, it is possible to compare the concentration of industries in the US and Mexico or that of high-tech and low-tech industries.

We calculate the raw concentration measure G, the Herfindahl index H, and the EG index r for 18 NIC 2-digit Indian industries. The results show that jute textile, beverages, leather and leather products, miscellaneous food products, wood and wood products, textile products, and wool and silk products have very high levels of local concentration, whereas non-metallic mineral, transport equipment and parts, machinery other than transport, electronic and electrical products, electronic and electrical machinery and parts and apparatus, rubber and petroleum and coal products, metal, and paper and paper products are hardly localized. The results indicate that more resource intensive industries tend to be more locally concentrated. Overall, spatial industrial distribution patterns in India resemble the concentration patterns of the US manufacturing industries that Ellison and Glaeser investigated. These summary findings are replicated when we map (in Figure 4.2) the distribution of firms for one highly concentrated sector (leather) and one far less concentrated sector (electronics).

We then examine labour productivity in rural, non-metro urban, and metro-urban areas. A simple comparison of productivity does not prove any causal relationship between economic geography and productivity differences. It is, however, meaningful since it can highlight important characteristics of firms located in different areas, which might result from location choices. Table 4.3 illustrates that there is a noticeable difference in labour productivity among firms in rural, non-metro urban, and metro-urban areas. Firms in large urban areas are substantially more productive than those in rural areas. The difference might be an outcome of economic geography, firm location choice, or both.

RESULTS FROM ECONOMETRIC ANALYSIS

The econometric analysis was carried out following the model specification discussed in an appendix to this chapter. The technical details of the model are available there. Before we begin discussing the results of the analysis, a few words on the significance of firm size will be useful. It is quite likely that due to heterogeneity in technology use, production efficiency, and managerial capacity among firms of different sizes, it may be limiting to group all firms in the same estimation process. Further, the benefits of location specific characteristics may be better accrued by smaller firms, which are relatively more dependent on access to buyers and suppliers, availability of ancillary services, inter firm non-technological externalities, and high quality infrastructure. In contrast, larger firms may be in a better position to internalize the production of various intermediate goods, self-provide infrastructure, and stock higher inventories. As a result, they are relatively less dependent on location based amenities and characteristics. To make allowances for this heterogeneity, and test the impact of economic geography across firms of different sizes, and find out if in fact there are differences in production costs, we classify firms into three categories: small, medium, and large. Small firms are defined as those with less than 50 employees, medium sized firms have between 50 and 99 employees, and large firms have 100 or more employees. The number of firms by size category is reported in Table 4.4.

Summary results for the estimated cost functions are reported in Tables 4.5 and 4.6. Table 4.5 provides results for the conventional inputs (capital, labour, energy and materials) and Table 4.6 provides estimates for the economic geography variables.[5] We present these separately as the economic geography variables are external effects, not directly included in the firm's cost structure. In these tables, we

Table 4.4: Number of establishments

Industry	Small (0-49)	Medium (50-99)	Large (100+)	Total
Food and beverages	1821	685	1498	4004
Textiles	1292	406	1621	3319
Leather	227	72	144	443
Printing and publishing	663	148	214	1025
Chemicals	1549	349	875	2773
Metals	1374	291	621	2286
Mechanical machinery	806	159	318	1283
Electrical and electronics	711	165	377	1253
Total	8443	2275	5668	16386

Source: Calculated from Annual Survey of Indian Industries, 1998–9.

provide results for the industry in general, followed by specific parameter estimates for small, medium, and large firms. From Table 4.5, it is quite clear that increase in factor prices translates into higher overall costs at the firm level.

Table 4.6 summarizes the impact of the economic geography factors on the cost structure (or profitability) at the level of the firm. The estimates in Table 4.6 are the cost elasticities of these variables, as defined in equation (14) in the appendix. There are four sets of location/economic geography variables in the analysis:

(1) access to markets (Access)
(2) own-industry concentration (Emp)
(3) buyer–supplier or input–output linkages (IO link)
(4) local economic diversity (Diversity)

The results for each industry sector are provided in four parts. The first column has industry wide cost elasticities. These are followed by estimates for small, medium, and large firms respectively. As we see from the results, sorting by firm size helps us identify particular types of firms, which are likely to benefit more from location based characteristics. In general, the cost elasiticities show that there is considerable heterogeneity in the impact of location characteristics on costs incurred at the firm level. This heterogeneity is not limited to the overall effects across industries, but also includes differences across firms of different sizes and by sources of agglomeration economies.

Table 4.5: Cost elasticities of production factors

	Capital				Labour				Energy				Material			
	Overall	Small (0–49)	Medium (50–99)	Large (100+)	Overall	Small (0–49)	Medium (50–99)	Large (100+)	Overall	Small (0–49)	Medium (50–99)	Large (100+)	Overall	Small (0–49)	Medium (50–99)	Large (100+)
Food and beverages	0.019	0.000	0.000	0.062	0.048	0.004	0.024	0.096	0.128	0.265	0.143	0.028	0.844	0.735	0.860	0.840
Textiles	0.051	0.013	0.021	0.087	0.096	0.029	0.065	0.184	0.141	0.133	0.120	–0.055	0.776	0.890	0.830	0.848
Leather	0.028	0.006			0.079	0.049			0.055	0.182			0.861	0.799		
Printing and publishing	0.082	0.009		0.129	0.139	0.053		0.222	0.318	0.206		0.321	0.692	0.788		0.726
Chemicals	0.069	0.027	0.089	0.103	0.106	0.056	0.124	0.204	0.242	0.263	0.114	0.103	0.716	0.776	0.724	0.683
Metals	0.035	0.004	0.112	0.054	0.092	0.055	0.131	0.134	0.280	0.243	–0.027	0.141	0.618	0.765	0.771	0.804
Mechanical machinery	0.037	0.026		0.084	0.144	0.112		0.192	0.197	0.241		0.096	0.746	0.757		0.689
Electrical and electronics	0.084	0.032		0.169	0.113	0.035		0.182	0.224	0.362		–1.047	0.740	0.650		0.600

Source: Authors' calculations.
Note: Coefficients in bold are significant at 1 per cent, coefficients underlined are significant at 5 per cent.

Table 4.6: Cost elasticities of economic geography variables

	Access				Emp				IO Link				Diversity			
	Overall	Small (0–49)	Medium (50–99)	Large (100+)	Overall	Small (0–49)	Medium (50–99)	Large (100+)	Overall	Small (0–49)	Medium (50–99)	Large (100+)	Overall	Small (0–49)	Medium (50–99)	Large (100+)
Food and beverages	−0.002	−0.001	−0.001	0.000	**0.016**	0.000	0.002	0.011	0.006	<u>−0.001</u>	−0.002	**0.024**	**−0.075**	0.000	−0.012	**−0.067**
Textiles	0.004	**0.023**	0.008	−0.022	<u>0.016</u>	0.004	<u>−0.033</u>	−0.001	−0.002	−0.005	**0.037**	0.011	**−0.102**	<u>−0.121</u>	**−0.210**	−0.005
Leather	**0.072**	−0.017			**0.025**	0.005			−0.014	0.010			−0.023	−0.172		
Printing and publishing	0.022	0.005		0.032	0.000	<u>0.012</u>		0.034	−0.009	−0.005		−0.092	−0.062	**−0.248**		0.0516
Chemicals	−0.016	<u>−0.024</u>	<u>−0.056</u>	−0.044	**0.021**	<u>0.021</u>	0.059	**0.049**	0.000	0.003	−0.041	−0.012	**−0.076**	**−0.457**	0.042	0.250
Metals	<u>−0.017</u>	<u>−0.008</u>	**0.163**	−0.012	0.003	0.004	<u>0.137</u>	−0.036	<u>−0.012</u>	0.000	**−0.177**	0.033	0.003	**−0.163**	0.603	0.039
Mechanical machinery	**−0.047**	−0.016		**0.091**	0.000	0.006	−0.018		−0.001	0.000	−0.026		−0.007	−0.042		0.167
Electrical and electronics	0.008	−0.009		0.035	**0.019**	<u>0.038</u>	0.035		−0.004	0.004		**0.378**	−0.162	**−0.835**		**−2.355**

Source: Authors' calculations.

Note: Coefficients in bold are significant at 1 per cent, coefficients underlined are significant at 5 per cent.

We start by describing the impact of access to markets. Market access, measured by transport network quality and urban population, measures effective demand for a firm's products and the ease with which it can reach buyers and suppliers. Locating in a region with good access to markets is likely to reduce the cost of intermediate inputs as well as increase demand for the firm's products. This will provide the entrepreneur with incentives to increase the scale of production and invest in cost reducing technologies. The industry wide results for market access suggest that the net cost reducing impact of market access is not significant in most industry sectors. The estimated cost elasticities are negative and statistically significant for two industry sectors—metals and mechanical machinery—the elasticity values are insignificant for other sectors. For example, in mechanical machinery, the coefficient of -0.046 means that a 10 per cent improvement in market access will be associated with an approximately 0.5 per cent reduction in overall costs at the firm level. We get a counter-intuitive result for the leather industry, where the cost elasticity is positive and significant.

For small firms, however, the estimated elasticities are generally negative, indicating benefits from improved market access. The estimates are statistically significant at the 5 per cent level for only two industry sectors: chemicals and metals. We also find a positive and significant estimate for the textiles industry, suggesting that there are costs associated with higher market access. Most of the estimates for medium and large industries are not statistically significant.

Following market access, we discuss results for own-industry concentration, which is measured as the sum of employment in the particular industry in the region. As in the case of market access, the reported estimates are elasticities. The industry wide estimates suggest that there are no net benefits of being located near own-industry concentrations. All the estimated elasticities are positive, which suggests that costs increase if firms locate in regions with high concentrations of the same industry. These coefficients are statistically significant at the 1 per cent level for four sectors and significant at 5 per cent for one industry sector. To examine if industry wide results are artifacts of aggregation, it is useful to look at the results by firm size. We find that even when disaggregated by firm size, own-industry concentration systematically provides either no net benefits and, in some instances, actually increases costs at the firm level.

The findings for input–output linkages show that for most industry sectors, proximity to buyers and suppliers potentially reduces costs at the firm level. While the estimated elasticities are negative for six sectors, it is only statistically significant at the 5 per cent level for the metals industry. The coefficient of −0.01 means that a 10 per cent increase in the strength of buyer–supplier linkages is associated with firm level cost reductions of 0.1 per cent. In other words, doubling the strength of buyer–supplier linkages is associated with a 1 per cent reduction in firm level production costs. When we look at the elasticities for small firms, we find that the estimates are insignificant in most cases. For medium size firms, the elasticity is negative and significant for the metals sector. The coefficient of 0.17 means that a doubling of IO linkages is associated with a 17 per cent reduction in firm level costs. This effect is considerably stronger than the other estimates, where the cost elasticities rarely exceed 5 per cent. For large firms, we find that costs increase for food and beverages and for electrical/electronics, when firms are located in regions with relatively higher buyer–supplier linkages.

The estimates for local economic diversity indicate that there are considerable cost reducing benefits from being located in an industrially diverse region. The industry wide estimates are negative for all sectors, and significant at the 1 per cent level for the food and beverages and textiles sectors. The coefficient of −0.10 for textiles means that doubling of the region's economic diversity will reduce firm level costs by 10 per cent. The results are even stronger for small firms. The estimated elasticities are negative for all industry sectors, and statistically significant for five sectors. What is really striking is the magnitude of these effects. For example, the estimated cost elasticity for electrical/electronics is 83 per cent and for chemicals it is 46 per cent. These estimates clearly suggest that there are considerable benefits of being located in a diverse economic region. The results for medium and larger firms, however, do not show similar benefits for location in diverse economic regions. The cost reducing effects of being located in a diverse region are greater for small firms because they can rely on location based externalities to a larger extent than medium and big firms. The benefits come from better opportunities for subcontracting, access to a general pool of skilled labour, and access to business services such as banking, advertising, and legal services. In addition to these pecuniary externalities, there are potential technological externalities from knowledge transfer across industries. Larger firms, being more vertically integrated and with higher fixed costs, are not likely to benefit from these externalities.[6]

In general, we find that the regional economic geography has a reasonable degree of impact on the cost structure of firms. The sources and the magnitudes of these impacts vary considerably across industry sectors. The only major source of benefits that is likely to influence location choice at the margin is the location's industrial diversity. This is further likely to be the case for small firms. The magnitude of the other effects are so small (elasticity values less than 5 per cent), that they are unlikely to influence firm location choices.

Results showing the effects of the economic geography factors on demand for traditional inputs are presented in Table 4.7. The estimated values are elasticities of substitution for input demands with respect to agglomeration factors, based on the specification in equation (16) in the appendix. Briefly, the following points may be highlighted:

(1) In general, economic geography factors have negligible substitution effects on capital. In most cases, coefficient estimates are negative, which implies cost-saving effects of economic geography factors, but are statistically insignificant. The only exceptions are the chemicals sector and the electrical/electronics sector where higher market access, IO links, and diversity lower the capital requirements, especially for small firms. For instance, doubling market access or IO linkages will reduce overall capital demand by 24.8 per cent and 17.7 per cent respectively. The fact that capital substitution effects are negligible is not surprising, because we cannot expect capital intensity or the cost of borrowing to vary over space.

(2) Labour requirements are consistently lowered with higher measures for economic geography variables (with the general exception of diversity) for the textile, printing/publishing, chemicals, machinery, and electrical/electronics sectors. These effects are most consistent for small firms, though the largest substitution effects are for large firms. Indeed, small firms are

Table 4.7: Input demand substitution

	Access				Emp				IO Link				Diversity			
	Overall	Small (0–49)	Medium (50–99)	Large (100+)	Overall	Small (0–49)	Medium (50–99)	Large (100+)	Overall	Small (0–49)	Medium (50–99)	Large (100+)	Overall	Small (0–49)	Medium (50–99)	Large (100+)
Food and Beverages																
Capital	−0.050	−0.041	3.670	−0.004	−0.080	0.018	0.119	−2.434	−0.056	−0.019	0.952	0.882	0.249	0.049	−4.751	1.791
Labour	−0.099	−0.056	0.706	0.148	0.000	0.024	0.121	0.173	−0.016	0.017	−0.012	0.113	−0.227	−0.005	−0.767	−0.373
Energy	0.559	0.873	−0.338	0.425	−0.717	0.966	−0.356	0.459	−0.076	0.763	−0.265	0.596	0.508	1.075	−0.334	0.322
Material	−0.037	−0.932	0.342	−0.555	0.057	−0.895	0.376	−0.513	0.012	−0.977	0.266	−0.459	−0.098	−0.852	0.330	−0.594
Textiles																
Capital	−0.484	0.073	−0.012	0.300	−0.078	0.021	0.003	0.020	−0.056	0.121	0.147	−0.062	0.386	−0.008	−0.092	−0.095
Labour	−0.224	−0.228	−0.241	−0.560	−0.096	−0.209	−0.258	−0.454	−0.016	−0.349	−0.164	−0.482	−0.267	−0.403	−0.385	−0.457
Energy	0.716	0.696	1.931	1.550	0.773	0.595	1.792	1.610	−0.076	0.901	1.999	1.596	0.615	0.703	1.678	1.622
Material	−0.548	−0.497	−1.683	−1.132	−0.516	−0.546	−1.794	−1.089	0.012	−0.430	−1.639	−1.092	−0.658	−0.575	−1.936	−1.089
Leather																
Capital	−0.012	−0.157			−0.010	−0.585			0.013	0.608			0.018	−0.050		
Labour	−0.050	0.153			−0.067	−0.020			−0.089	−0.049			−0.137	−0.238		
Energy	−1.161	0.101			−1.125	−0.042			−1.079	−0.004			−1.028	−0.241		
Material	0.957	−0.013			1.170	0.048			1.369	0.059			1.564	−0.120		
Printing and Publishing																
Capital	0.388	−0.570		0.271	0.188	−0.400	0.350		0.052	1.177		−0.130	−0.449	−0.423		−0.224
Labour	−0.373	−0.134		−0.595	−0.337	−0.129	−0.540		−0.404	−0.016		−0.639	−0.422	−0.384		−0.493
Energy	0.338	0.585		0.620	0.180	0.730	0.398		−0.158	0.532		0.978	−0.131	0.469		0.963
Material	0.307	−0.491		0.178	0.326	−0.440	−0.063		0.749	−0.522		0.349	0.476	−0.703		0.271

Table 4.7 (contd)

Chemicals																
Capital	−0.248	−0.085	−0.129	−0.331	−0.147	−0.062	0.077	−0.085	−0.177	−0.085	−0.151	−0.178	−0.187	−0.531	0.029	0.210
Labour	−0.241	−0.536	−0.500	−0.548	−0.038	−0.477	−0.341	−0.281	−0.203	−0.504	−0.480	−0.480	−0.246	−0.963	−0.394	−0.149
Energy	0.487	0.259	0.785	−0.097	−0.332	0.659	1.004	1.039	0.593	0.040	0.795	−0.109	−0.310	0.301	1.090	0.778
Material	0.294	0.225	−0.530	0.313	0.505	0.374	−0.290	0.584	0.193	0.144	−0.475	0.124	0.307	−0.098	−0.270	0.551
Metals																
Capital	0.010	0.108	−0.163	0.139	−0.113	0.455	−0.153	−0.997	0.051	−0.417	−0.483	0.580	0.055	−0.259	0.352	0.462
Labour	0.057	−0.181	0.074	−0.136	0.061	−0.225	−0.001	−0.179	−0.010	−0.106	−0.412	−0.025	−0.042	−0.355	0.426	−0.088
Energy	1.850	0.527	−0.844	0.840	1.940	1.138	−0.961	0.796	1.899	−0.592	−0.917	1.154	2.029	0.118	−0.166	1.087
Material	−2.000	−0.093	1.292	−0.814	−1.934	−0.020	1.202	−0.852	−1.969	−0.210	1.448	−0.657	−1.877	−0.284	2.158	−0.682
Mechanical Machinery																
Capital	−0.025	−0.009		−0.032	0.071	0.022		−0.053	0.163	0.132		−0.048	0.247	0.156		0.169
Labour	−0.204	−0.668		0.645	−0.116	−0.630		0.392	−0.185	−0.658		0.005	−0.175	−0.679		−0.231
Energy	1.273	1.004		−0.916	1.429	1.206		−0.988	1.435	1.181		−0.904	1.587	1.347		−0.723
Material	−1.419	−0.639		0.748	−1.303	−0.493		0.766	−1.300	−0.524		1.010	−1.210	−0.411		1.259
Electrical/Electronics																
Capital	−0.235	−0.229		−0.531	−0.131	−0.131		−0.488	−0.215	−0.199		−0.158	−0.335	−0.982		−2.865
Labour	−0.320	−0.777		−0.441	−0.210	−0.714		−0.361	−0.307	−0.761		−0.095	−0.430	−1.592		−2.803
Energy	0.225	1.718		−0.784	0.023	1.709		−0.444	0.093	1.676		−2.819	−0.260	0.843		−2.888
Material	0.681	−0.617		1.100	0.564	−0.643		1.256	0.551	−0.673		1.469	0.396	−1.523		−1.158

Source: Authors' calculations.
Note: Coefficients in bold are significant at 1 per cent, coefficients underlined are significant at 5 per cent.

more likely to benefit from economic geography factors by tapping into external economies scale that such factors bring about. We believe that these effects are related to access to skilled labour; skilled/productive labour is likely to be available in areas with better access, high own-industry concentration, diversity, etc. Hence, it is possible to use a smaller workforce in places with superior economic geography.

(3) Energy requirements, on the contrary, are increased with higher values for economic geography variables. The coefficients are consistently significant in the textiles and machinery sectors, and generally significant in the food and metals sectors. This effect is probably related to the Byzantine energy pricing methods used by Indian state electricity boards. In most cases, the cross-subsidy systems punish urban industrial consumers to reward agricultural and residential consumers. As a result, energy costs are higher in urban/metropolitan areas even if energy requirements remain the same.

(4) The patterns of substitution for materials is inconsistent between sectors and firm sizes. The only consistently significant substitution effects are in the textiles sector, but even there we can see some variation (different signs) for different methods of firm aggregation. It is not possible to find general explanations for what appears to be a random pattern.

Although the results only partially show that economic geography factors affect traditional input factor demands in a consistent way, this does not contradict previous findings. The elasticities of substitution of externality variables with respect to input demands are often inconsistent, especially for materials. For instance, Bernstein (1988) and Bernstein and Nadiri's (1988) studies on R&D spillovers showed that the elasticities of substitution of R&D spillovers with respect to traditional inputs do not have consistent patterns among different industries.

CONCLUDING COMMENTS

In conclusion, we would like to highlight three points. First, the analytic strategy and empirical specifications used here are original, comprehensive, and generalizable. Though our work is motivated by development issues, and the findings contribute to the literatures on urban, regional, and industrial development, the methodology developed here is not limited to the analysis of just India or other developing countries only. This strategy can be applied to most firm-level examinations of location decisions in any country. In this regard, this approach is a significant advance in the spatial analysis of industrialization, and specially the large and growing field focusing on externalities, clustering, and increasing returns.

Second, the principal finding—that industrial diversity (that is, the local presence of a mix of industries) provides significant cost savings for individual firms, and is the only economic geographic variable to do so consistently—raises serious questions about the validity of much theorizing on localization economies. Our analysis shows that this cost saving is the most significant factor for firms of all sizes and in all sectors of the manufacturing industry. Other spatial factors that, in theory, have some productivity enhancing effects or cost benefits (such as local presence of own industry, local access to buyers and suppliers) are found to have little or no influence on profitability. In other words, localized external economies have few discernible cost benefits. Rather, generalized urbanization economies (manifested in local economic or industrial diversity) provide the agglomeration externalities that lead to industrial clustering in metropolitan and other urban areas. In the next chapter, we will show that this finding holds true at smaller scales, that is, within metropolitan regions.

Third, the policy implications of the findings are quite significant. Consider only the spatial policy issues. The findings on the traditional production inputs, especially those pertaining to energy costs, are important, but deserve a separate and detailed treatment. The validity of developing 'specialized clusters' in remote areas, as instruments to promote regional development in lagging or backward regions, must be questioned. Such approaches have been implemented with limited success historically (witness the rise and fall of the 'growth pole' concept), but have seen resurgence with the 'Porter style' competitive advantage analysis. In contrast, policies that encourage the creation and growth of mixed industrial districts are likely to be more successful than single industry concentrations. However, this is easier said than done, especially in remote or lagging regions. If location related cost advantages are not related to market access (whereby dispersed infrastructure investments, particularly in transportation, do not favour lagging regions), or localization economies, it is difficult to see how manufacturing industry can become the engine of growth in lagging regions. We return to this issue in the final chapter.

NOTES

1. By grouping firms into carefully defined sectors (rather than examining all manufacturing together), we can identify the differential impact of regional characteristics or geographic externalities across industries. For example, in comparison to food processing, which is closely linked to the traditional rural industrial base, industries such as machinery, metals, and computers and electronics are relatively footloose urban industries subject to considerable agglomeration economies.
2. NEG's approach bears a strong resemblance to Marshall (1890) and Weber (1929) in many ways. However, unlike its predecessors, new economic geographers place less emphasis on technology spillovers as a source of externalities than on labour pooling and specialized suppliers. Krugman (1991b) argued that externalities from technology spillovers are difficult to measure, and therefore, cannot be modelled. Instead, he argued that under increasing returns to scale and imperfect competition, pecuniary externalities have clear welfare effects due to the variety of market size effects (that is, each firm's monopoly power can affect the production function of other firms through buying and selling in the market) (Krugman 1993). By focusing on pecuniary externalities (or rent spillovers) rather than technology spillovers, NEG tries to focus the general discussion on externalities.
3. The total road length is determined using a geographic information system. Due to generalization at large cartographic scales, this represents a low estimate of the total length of all roads in the data set. Furthermore, the digital roads data are unlikely to include all roads in the country. According to Indian government figures, the total length of surfaced roads was about 1 million kilometres in 1991, of which 34,000 km are national highways and 128,000 km state highways (CMIE 1998). However, we assume that the data set contains all major links between urban centres and that the results are unlikely to be affected by omitted minor roads.
4. Goldar (1997) notes that factories are classified into industries according to their principal products. In some cases this causes reclassification of factories from one class to another in successive surveys, making inter-temporal comparisons difficult.
5. There are some cells in Tables 4.5 and 4.6 with no values. We do not report the estimated parameters in these cases as the number of observations (see

Table 4.4) is too few to allow any meaningful interpretation of the results—especially when the model estimates around 50 parameters. As a rule of thumb, we do not report results for estimations with less than 200 observations (firms).
6. While the estimated elasticity for large electrical/electronics firms is 235 per cent, it is likely that this result is a statistical artifact, and driven by some outliers.

APPENDIX

ECONOMETRIC SPECIFICATION AND DATA IMPUTATION

In this first part of the appendix, we present the econometric specification to test the effects of economic geography factors in explaining the location of economic activity. Our basic premise is that firms will locate in a particular location if profits exceed some critical level demanded by entrepreneurs. We estimate a cost function with a mix of micro-level factory data and economic geography variables, which may influence the cost structure of a production unit.

A traditional cost function for a firm i is (subscript i is dropped for simplicity and the equation numbering system is continued from the main text):

$$C = f(Y, w) \qquad (9)$$

where C is the total cost of production for firm i, Y is its total output, w is an n-dimensional vector of input prices. However, as discussed earlier, the economic geography, or the characteristics of the region where the firm is located, is also an important factor affecting the firm's cost structure. Such location-based advantages have clear implications for a firm's location decision as they create cost-saving externalities. We modify the basic cost function to include the influence of location-based externalities:

$$C_r = f(Y, w_r, A_r) \qquad (10)$$

where C_r is the total cost of a firm in region r, w_r is an input price vector for the firm in district r, and A is a m-dimensional vector of spatial externalities (that is economic geography or agglomeration variables such as access to markets, buyer–supplier networks, own-industry concentration) at location r.

The model has four conventional inputs: capital, labour, energy, and materials. Therefore, the total cost is the sum of the costs for all four inputs. With respect to agglomeration economies, it is assumed that there are four sources of agglomeration economies at the district level (described in the previous section) such that $A=\{A_1, A_2, A_3, A_4\}$, where A_1 is the market access measure, A_2 is the concentration of own-industry employment, A_3 is the strength of buyer–supplier linkages, and A_4 is the industrial diversity in the region.

Shephard's lemma produces the optimal cost minimizing factor demand function for input j corresponding to input prices as follows:

$$X_{jr} = \frac{\partial C_r}{\partial w_{jr}}(Y, w_r, A_r) \quad j = 1, 2, 3, 4, ..., n \tag{11}$$

where X_{jr} is the factor demand for j^{th} input of a firm in district r. It is clear that the firm's factor demand is determined by its output, factor prices, and location externalities. Therefore, the production equilibrium is defined by a series of equations derived from equation (9) and (10).

The empirical implementation of the above model is based on a translog functional form, which is a second-order approximation of any general cost function. Since there are four conventional inputs and four location externalities (agglomeration) variables, a translog cost function can be written as:

$$\ln C = \alpha_0 + \alpha_y \ln Y + \sum_j \alpha_j \ln w_j + \sum_l \alpha_l \ln A_1 +$$
$$1/2 \beta_{yy} (\ln Y)^2 + 1/2 \sum_j \sum_k \beta_{jk} \ln w_j \ln w_k +$$
$$\sum_j \beta_{jy} \ln Y \ln w_j + 1/2 \sum_l \sum_q \gamma_{lq} \ln A_l A_q +$$
$$\sum \sum \gamma_{jl} \ln w_j A_l + \sum_l \gamma_{ly} \ln Y \ln A_l$$
$$(j \neq k; l \neq q; j, k = 1, 2, 3, 4; l, q = 1, 2, 3, 4) \tag{12}$$

In addition, from equation (10), the cost share of input factor j can be written as

$$S_j = \alpha_j + \sum_k \alpha_{jk} \ln w_k + \beta_{jy} \ln Y + \sum_l \gamma_{jl} A_l$$
$$(k = 1, 2, 3, 4; l = 1, 2, 3) \tag{13}$$

Notice that the cost share equations of all factor inputs satisfy the adding up criterion, $\Sigma_j S_j = 1$. The 'adding up criterion' has important implications for model estimation. The system of cost share equations satisfies the 'adding up criteria' if

$$\sum_j \alpha_j = 1; \quad \sum_k \beta_{jk} = \sum_j \beta_{jk} = 0; \quad \sum_j \beta_{jy} = 0;$$
$$\sum_l \gamma_{jl} = \sum_j \gamma_{jl} = 0 \tag{14}$$

thereby, reducing the number of free parameters to be estimated.

The translog cost function can be directly estimated from equation (12). However, a joint estimation of equation (12) and (13) with restriction (14) significantly improves the efficiency of the model. The final model estimated includes two additional dummy variables that identify locational characteristics that may not be captured by agglomeration variables. Locations are categorized as rural, non–metro urban (D_1), and metro-urban (D_2), and rural location is used as a reference category.

The impact of the economic geography factors on the cost structure (or profitability) of the firm can be evaluated by deriving the elasticity of costs with respect to the economic geography variables. From equation (12) the cost elasticities are:

$$\frac{\partial C}{\partial A_l} = \alpha_l + \Sigma_j \gamma_{jl} \ln w_j + \Sigma_q \gamma_{lq} \ln A_q + \gamma_{ly} \ln Y \qquad (15)$$

In addition to direct impact on the cost structure, these location specific externalities also influence factor demand. The impact of these variables on input demand can be derived from the cost share equations. Note that the cost share for input j, S_j, can be written as $w_j v_j / C$, where w_j is factor price of input j, v_j is the quantity demanded of input j, and C is total cost. That is,

$$v_j = \frac{C}{w_j} S_j \text{ and } \ln v_j = \ln C + \ln S_j - \ln w_j \qquad (16)$$

Therefore, the elasticities of input demands with respect to agglomeration factors A_l is

$$\frac{\partial \ln v_j}{\partial \ln A_l} = \frac{\partial C}{\partial A_l} + \frac{\gamma_{jl}}{A_l} \qquad (17)$$

The empirical analysis is conducted by jointly estimating equations (12) and (13) as a system, using an iterative seemingly unrelated regression (ITSUR) procedure. The underlying system is nonlinear, and is primarily derived from the structure of the input demands, as represented in equation (12). The ITSUR procedure estimates the parameters of the system, accounting for heteroscedasticity, and contemporaneous correlation in the errors across equations. As the cost shares sum to unity, $n-1$ share equations are estimated (where n is the number of production factors). The ITSUR estimates are asymptotically equivalent to maximum likelihood estimates and are invariant to the omitted share equation (Greene 1997). All estimations were carried out with the MODEL procedure of the SAS system.

DATA IMPUTATION

The 1999 ASI data used for estimation have a significant number of incomplete cases. Many firms did not report their capital, output, depreciation, and other related input price information. Even when there are reported values, some of them are not consistent (for example 0 capital when capital depreciation is reported positive). Missing or inconsistent data can be a serious problem when such data points are not completely random.

To take into account the limitations arising from the less than perfect ASI data, we first adopted the following set of rules to clean the data, and then imputed missing values in the cleaned data using SAS MI procedure. First, cases that are missing too much vital information (for example input, output, capital, and employment) are deleted (only 78 cases were deleted from this step). Second, when the value for plant and machinery depreciation is positive and the size of employment is greater than 10, but the closing value of capital is reported 0, capital is converted to missing. Lastly, when capital is missing, but its depreciation value is 0, depreciation is converted to missing because it is likely that newly imputed values for capital will be positive which implies positive depreciation of capital.

The easiest and probably the most frequently used methods to handle missing data points are casewise data deletion and mean substitution. If a case has any missing values, the entire record can be deleted or missing points can be substituted by mean values. However, Roth (1994) compared different approaches often used in empirical research and concluded casewise data deletion and mean substitution are inferior to maximum-likelihood based methods such as multiple imputation.

To resolve the issue of missing data, we introduce a multiple imputation technique developed by Rubin (1978, 1987) and others. The multiple imputations usually generate five to ten complete data sets by filling in gaps in existing data with proper raw values. Raw values are drawn from their predicted distribution based on the observed ones. Then each complete data set can be analysed by common statistical methods (for example regression). After conducting identical analysis multiple times, the results drawing upon imputed data sets are combined into one summary set of parameters.

We generated five complete data sets and used mean values to impute missing cases. The imputed values were evaluated again to check their consistency. When imputed values were unreasonably small or large, we converted them back to missing and imputed again. The imputation procedure was repeated three times.

These plant level data are supplemented by district and metropolitan area level demographic and amenities data from the 1991 Census of India and detailed information on the availability and quality of transport infrastructure linking urban areas. The plant level data have been combined with district level indicators such as concentration of industry in the district, urban population density, and potential access to urban markets.

5

Industrial Clusters within Metropolitan Regions

In previous chapters, we have described and analysed the distribution of industry at the macro (states) and meso (districts) scales. In this chapter, we turn to the micro scale, that is, the distribution of industry within metropolitan areas. There are several reasons why this scale of analysis is important.

(1) First, it is only at this scale that clustering takes place; therefore, this is the scale at which localization economies, one of the driving forces behind the formation of agglomerations, can be expected to be realized. A district may have twenty or thirty firms in the textiles sector, but unless they are located in close proximity, it is theoretically possible for them to be stand-alone units; hence, it is incorrect to assume that they are clustered and that they realize localization economies.

(2) Second, one of the knotty problems in economic geography is the fuzziness of the concepts of localization and urbanization economies (as discussed in Chapter 1). Where does one end and the other begin? This conceptual problem is especially acute in metropolitan areas which have both clusters of similar firms in the proximate area and arrays of supporting service firms in the general area.

(3) Third, governments at all levels—national, sub-national, and local—have zeroed in on the possibility that localization or clustering benefits production and have created a host of policies designed to take advantage of this possibility. These land-use policies have various names—industrial zones, enterprise zones, export-processing zones, free trade zones, technology parks, etc.—but share common features: Firms are

encouraged to locate in narrowly delineated areas well furbished with infrastructure (and, sometimes, tax incentives), usually inside metropolitan regions but at some distance from the congested core. In fact, policies for intentional clustering may at this point be the most important of the direct industrial policies used by sub-national and local states.

As a result, industrial clusters have re-emerged as important objects of research and policy analysis. In this chapter, we present and analyse micro-level industry location data (at the scale of the postal pin code) for three metropolises: Mumbai, Calcutta, and Chennai. We show a picture that has never been seen in the Indian context (there simply is no other study at this spatial scale), and at the same time we interrogate one of the critical questions in economic geography: do localization economies matter?

Since data at this level of geographic detail have never been analysed in India, we begin by asking the basic empirical questions for specific industries:

(1) Where do industries locate within a metropolitan area?
(2) Do industries cluster?
(3) Do different industrial sectors have different patterns of location/clustering?
(4) Can these patterns be understood with reference to industry or firm characteristics?

We then consider questions relating to the spatial relationships between the different industry sectors; specifically, we ask: do industries co-locate or co-cluster? Through all this, our goal is to answer questions on location theory: What is the role of localization economies in cluster formation in metropolises? To what extent do these derive from inter-firm transactions such as collaboration and information sharing, or intra-industry transactions such as through shared buyer–supplier networks, intermediate goods, and specialized labour pools. What is the role of urbanization economies as manifested in regional industrial diversity, general (as opposed to specialized) labour pools, and buyers and suppliers and knowledge transfers in the region as opposed to the neighbourhood?

In the following pages, we first provide the necessary background: the theory of industrial clustering, the data used for the analysis, and cluster measurement methods. Next, we test eight industrial sectors (the ones used in Chapter 4, that is food/beverages, textiles, leather, printing/publishing, chemicals, metals, machinery, electrical/electronics) for evidence of global and local clustering (explained later), and distinguish between and test for co-clustering and co-location of industries (also explained later). The results suggest a temporal model of industry location in mixed rather than specialized industrial districts. There is little evidence that localization economies drive industrial location decisions. In the concluding section, we detail the argument that land-use policies—segregationist/environmental policies, the absence of exit and land-use change policies, and activist industrial promotion policies—are the key influences on the intra-metropolitan spatial distribution of industry.

LOCALIZATION ECONOMIES AND CLUSTERING

Why do Industries Cluster?

We know from Chapter 1 that the links between industry clustering and localization economies were identified early by Alfred Marshall (1919). He also identified the primary reasons why firms benefit from localization: namely, the availability of common buyers and suppliers, the formation of a specialized/skilled labour pool, and the informal transfer of knowledge (on trade secrets, production processes, market agents, etc.). Krugman's work in economic geography (Krugman 1991a, 1996) and Porter's work in business economics (Porter 1990, 1996) have drawn the interest of economists to the idea of 'increasing returns' to proximity in the form of clusters (see Fujita, Krugman, and Venables 1999). Meanwhile, a tradition of studying the locational aspects of economic activity exists in several academic disciplines, especially regional science and geography (for a recent survey, see Walker 2000). The literature on industrial clustering and its causes (localization and/or urbanization economies, proximity to other firms and/or consumers) and effects (economic growth, unbalanced development, regional inequality, global industrial restructuring) continues to proliferate in journals of geography, economics, planning, and development (see Chakravorty 2000; Henderson and Kuncuro 1996; Lee 1989; Nadvi and Schmitz 1999; World Bank 1999; and others for empirical work in developing nations). Let us begin by understanding the concept more clearly.

Clustering is a term describing a phenomenon in which events or artifacts are not randomly distributed over space, but tend to be organized into proximate groups. Industrial clustering is a process that has been observed from the beginning of industrialization. From the cotton mills of Lancashire and automobile manufacturing in Detroit, to the textile mills of Ahmedabad and Mumbai and the tanneries of Calcutta and Arcot, even the casual observer can visually identify industry clusters. It seems obvious that competing firms in the same industry derive some benefit from locating in proximity to each other. The benefits that are external to the firm and accrue to similar firms in proximity are called the economies of localization. Now, these typically are not the only firms in the immediate region. There are usually other factories, producing other and similar goods, distributed through other and similar channels, for other and similar markets. These other firms, and their employees, and the service workers who provide food, education, transportation and health care for all these employees and their families, comprise, typically an urban area. All the firms that benefit from being in the urban area, regardless of whether or not there are other similar firms in the area, derive economies of urbanization from their location choice.

To put it in another way, at the firm level, it is expected that the size and number of firms (that is the competitive structure) will influence internal returns to scale. In particular, as demand for a firm's goods and services increases (say, due to improved access to consumer markets), the entrepreneur has an incentive to increase the scale of production by restructuring the production process through the use of specialized workers and investing in cost-reducing technologies (Lall et al. 2004a). At the industry level, we expect to see quantifiable localized benefits of clustering which accrue to all firms in a given industry or in a set of inter-related industries. Productivity is likely to be higher in regions where an industry is more spatially concentrated due to the increased potential of knowledge spillovers and dense buyer–supplier networks, access to a specialized labour pool, and opportunities for efficient subcontracting. Finally at the metropolitan area level, economies of scale result not from the size of a specific industry or market but from the overall size, diversity, and spatial configuration of the metropolitan area. These economies of urbanization include access to specialized financial and professional services, availability of a large labour pool with multiple specializations, inter-industry information transfers, and the availability of less costly general infrastructure (see Parr 2002). At the inter-regional scale, these gains are expected to lead to industry concentration in metropolitan and other leading urban regions (as a result of urbanization economies); at the metropolitan scale the gains from localization economies are expected to lead to the creation of local industrial clusters.

These typically unquantified agglomeration economies are one set of inputs into the location decision of a firm. There are other significant factors that a firm facing a location decision must consider. The two most important of these additional factors (especially in developing countries) are the availability of infrastructure, and the regulatory framework, both arenas where the state is the key player. The state not only sets the rules of market entry and participation, but is also the primary, often the sole provider of physical and social infrastructure, and is often directly active in the production process. In India, in its efforts to capture the 'commanding heights of the economy,' the state invested heavily in capital-intensive industry, such as integrated steel and power plants (detailed in Chapter 2). It was so successful in its efforts that as recently as the late 1990's, nine of the top ten and twenty of the top twenty-five corporations in India were public sector units (Nayar 1998). At the local level, the state's regulatory role goes beyond setting the rules of market participation; by being the single largest owner of land, by having the police and taking powers to acquire necessary land, and by being the final arbitrator on land-use decisions, the state, as we shall show later, has a very strong influence on industrial location decisions within metropolitan areas.

WHAT DO WE NOT KNOW?

Some fundamental issues remain unresolved. In this chapter, we address what we think is the most important of these issues, namely, the relative importance of localization versus urbanization economies in cluster formation in Indian metropolises. The analytical problems begin with the fact that there is little agreement on what a cluster is in empirical terms: How many firms constitute a cluster? How many workers should a cluster have? What is the geographic area over which we count these numbers? How close is close enough, how far is too far? According to Martin and Sunley (2003: 12) 'to use the term [cluster] to refer to any spatial scale is stretching the concept to the limits of credulity, and assumes that "clustering processes" are scale-independent.' It seems to us that the constituent elements of agglomeration economies—which are localization and urbanization economies—conceptually account for most scale issues. Localization economies refer to features that are very proximate, being in the neighbourhood, while urbanization economies refer to features that are proximate, but less so, being in the region rather than in the neighbourhood. This is far from precise, because neighbourhood and region remain undefined. Later we will be more specific.

The analytical problems continue with the implicit assumption in the clustering literature that industry location decisions are made in flexible, if not unfettered, land markets; that decision makers at the firm level choose locations from a multitude of options. This is a critical assumption, because a firm's location decision is fundamentally a choice in the land market. We know that even in strongly market oriented economies (such as the United States) there are land market constraints like zoning. In fact, clustering policies a la Porter rely on the use of state power to intervene in land markets.

We argue that land market rigidities exist everywhere; in the specific case of Indian metropolises, these rigidities seriously constrain location choices for firms, so much so that localization economies practically do not matter when the location choice is made. The rigidities in the land market arise from state actions, specifically industrial location policy, environmental policy on segregating or buffering out polluting industry, and land-use policy on land-use change. This does not invalidate the possibility of realizing localization economies after the location decision; in fact, it is quite possible that the land market rigidities aid clustering, which may lead to localization economies during the production phase. However, the evidence indicates that most manufacturing firms have relatively few location choices when they enter the market, and probably the only external economies they can surely realize are general urbanization economies. At the point of the location decision, the region matters more than the neighbourhood.

NOTES ON METHODOLOGY

Data Issues

In order to undertake sub-metropolitan level analysis, it is necessary to have spatially disaggregated data. As discussed in Chapter 2, Indian industrial data are collected by the CSO and disseminated as the ASI. In the late 1990s, the ASI data were first released at the district level and then at the firm level. These data are collected from a survey undertaken by CSO on a sample taken from an industry sampling frame which includes every registered (or legal) industrial unit with at least ten workers. This sampling frame contains one record for each industrial unit and includes three critical pieces of information: the NIC code, the number of workers in the unit (the only piece of 'data' so to speak), and the street address, with, sometimes, a postal pin code (equivalent to US zip codes).

The last piece of information is the key to disaggregating the district data down to smaller enumeration units. We have access to the sampling frame for the whole country for the enumeration period 1998–9. On further examination, we found that while street addresses were generally available for all metropolitan areas, there were no base maps of streets to which these addresses could be matched. Hence, we had to rely on the pin code information, which, however, turned out to be erratically available. For some cities the pin codes were generally available or imputable, for other cities they simply were not available. We identified Mumbai, Calcutta, and Chennai as the three metropolises with enough information to begin geocoding the industry location data to the pin code level.

The pin code maps were acquired from a private sector firm in New Delhi (ML Infomap 1998). These maps have somewhat variable coverage. For Calcutta and Chennai, these maps cover the largest definition of their metropolitan areas. For Mumbai, the pin codes cover the district of Greater Bombay only; that is, the far northern and eastern suburban reaches of the Mumbai metropolitan area (in Thane and Raigad districts) are not covered. Even in the covered area, there appear to be some situations where adjoining pin codes have been merged. As a result, some data that are known to be in the metropolitan area could not be geocoded to pin codes. We have been able to achieve the following 'hit' rates, that is successfully geocoded factory records: in Mumbai 99.9 per cent, in Calcutta 97.1 per cent, in Chennai 97.5 per cent. Following Ratcliffe (2002), who argues that a 'hit rate' of 85 per cent is acceptable for most map-based analysis, we believe that these are acceptable levels of address matching. Finally we aggregated the firms into eight distinct and internally consistent sectors, as discussed in Chapter 4.

THE MEASUREMENT OF CLUSTERING

Spatial statistics are the most widely used tools for identifying and analysing spatial patterns (see Getis and Ord 1992 and Anselin 1995 for excellent discussions on the subject). In classifying spatial patterns, researchers are often interested in determining whether the distribution of activity is clustered, dispersed, or random. At this point, it is useful to differentiate between industrial 'clustering' and 'concentration'. Several devices to measure industry concentration have recently been operationalized. The so-called 'spatial Gini' and γ (from Ellison and Glaeser 1997) have become quite well known. These measures, and older ones such as the Location Quotient (discussed in Chapter 4 and later in this chapter), suffer from a common drawback, one that White (1983) termed the 'checkerboard problem', whereby these measures are not really spatial—any geographical arrangement of parcels (in this case pin codes) would yield the same measure of concentration. Hence 'concentration' has to be distinguished from 'clustering' where the latter is explicitly spatial; that is, geographical arrangements are incorporated in measures of clustering, but not in measures of concentration.

Clustering is best understood in the context of spatial autocorrelation (discussed earlier), a term that describes conditions where the attribute values being studied are correlated according to the geographic ordering of the objects. When the location of firms is spatially autocorrelated, it implies that the geographic distribution of economic activity is not random and is likely to be determined by some underlying political/economic/physical factors attributable to each geographical unit. Hence, strong positive spatial autocorrelations mean that the attribute values of adjacent geographical units are closely related.

One of the most popular measures of spatial autocorrelation is Moran's I. There are two types of Moran's I. The Global Moran is a measure describing the overall spatial relationship across all geographical units. This has been discussed extensively in Chapter 2. Recall, that for the Global Moran, only one value is derived for the entire study area. On the other hand, the Local Moran (often called Local Indicator of Spatial Association or LISA) is a measure that is designed to describe the heterogeneity of spatial association across different geographical units. The Global Moran can be thought of as a 'regional' measure, the Local Moran as a 'neighbourhood' measure. If neighbouring units have similar values over the entire study area, the statistic will show a strong positive spatial association. If dissimilar values are observed among neighbouring units, the statistic should indicate a strong negative spatial association. However, the magnitude of spatial association is not necessarily or usually uniform over the space. It is more likely to be heterogeneous according to local characteristics that influence the formation of spatial structure. Local Moran's I can be used as an indicator of heterogeneity in spatial association over geographical units and is defined as

$$I_i = z_i \Sigma_i w_{ij} z_j, \text{ where } z_i = (x_i - \bar{x}) / \delta$$

where, z_i is deviation from mean, and d is the standard deviation of x_i. Similar to global Moran, a high value of local Moran implies the association of similar values whereas a low value means the association of dissimilar values. It is important to note that local instabilities or clusters may exist in a given spatial distribution even when the distribution as a whole does not exhibit a statistically significant level of spatial autocorrelation.

This analytical method automatically answers a key question identified at the beginning: what is a neighbourhood, or over what geographical area do we calculate clusters? In most cases, a neighbourhood is defined as the sum of a sub-regional areal unit and its adjacent (or first order) neighbours. In this case, it is a pin code and its contiguous pin codes. It is possible to use other formulations of neighbourhood. We can use distance based measures (as we did in Chapter 2), whereby every areal unit whose centroid falls within a chosen distance of the centroid of a given areal unit becomes its neighbour. Also, we can use not only immediately adjacent neighbours, but second order neighbours, that is, immediate neighbours of immediate neighbours. These latter definitions would form larger neighbourhoods. There is no theoretical guidance on which of these measures is appropriate for which kind of analysis. We used both first-order neighbours and a distance cut-off definition of neighbourhoods. Only the contiguity-based (or first order) local Morans are reported here. The distance-based measure yields similar results.

THE STUDY AREAS

Mumbai (Bombay), Kolkata (Calcutta), and Chennai (Madras) are three of India's four largest metropolitan areas. All are colonial cities, created by the British originally as bases for trade (as ports) and which later became very important centres of regional administration (see Kosambi and Brush 1988 on their structure and morphology). Our study areas cover the district of Greater Mumbai (with 94 pin codes), metropolitan Calcutta (with 133 pin codes), and metropolitan Chennai (with 108 pin codes). The detailed geographies of these three metropolises identifying selected sub-metropolitan neighbourhoods are shown in Figures 5.1 through 5.3.

Mumbai is the sub-continental leviathan. The population of the metropolitan area is estimated to be over 18 million, making it, with Mexico City, the second largest urban agglomeration in the world. The metropolitan population has grown by over 45 per cent in each of the two preceding decades. Mumbai is the centre of the financial sector in India, home of the Reserve Bank of India and the majority of the state-owned and international banks and financial institutions. It is also a major industrial hub, one end of India's most dominant industrial region stretching north up to Ahmedabad in Gujarat state; this region has done particularly well in attracting new industrial investments after the 1991 structural reforms even while the district of Greater Mumbai has lost some share (see Chapter 2).

Calcutta was the most important colonial city in India, the seat of the British empire and its political centre till 1911, and the industrial centre of the nation till the early 1960s. The Calcutta metropolitan region is the second largest in the country, after Mumbai, with around 12.5 million people; till the 1991 census, it had been India's largest metropolis. The city and the region have seen a dramatic decline in industry in general; its specializations are in 'sunset' industries, specifically jute (textiles) and iron and steel. The average income of the state has declined from an all-India high in 1960 to around the

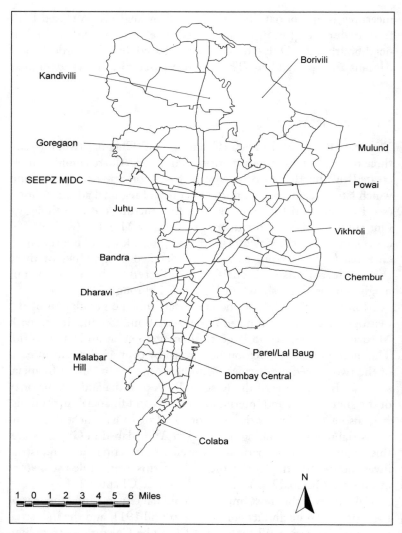

Figure 5.1: Greater Bombay and its sub-regions

Source: Digital base map acquired from ML Infomap (1998).

national average now. Since 1977, the state of West Bengal has been ruled by a coalition of communist and leftist parties which resisted the 1991 structural reforms and elements of which continue to be seen to be deeply suspicious of capitalism and globalization.

INDUSTRIAL CLUSTERS WITHIN METROPOLITAN REGIONS 153

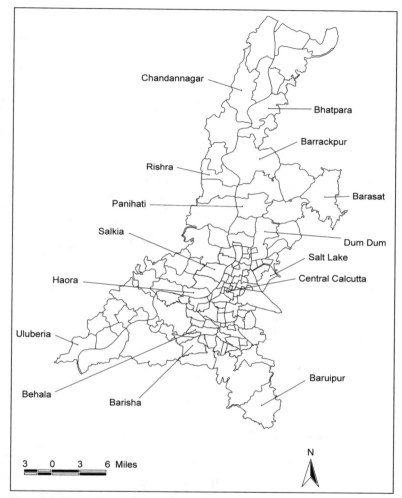

Figure 5.2: Metropolitan Calcutta and its sub-regions

Source: Digital base map acquired from ML Infomap (1998).

Chennai is the fourth largest metropolitan area in India (behind Mumbai, Calcutta, and Delhi). It was the pre-eminent city of south India for most of the twentieth century, till Bangalore, in the neighbouring state of Karnataka, emerged as a serious regional rival, primarily as a result of the growth of the information technology

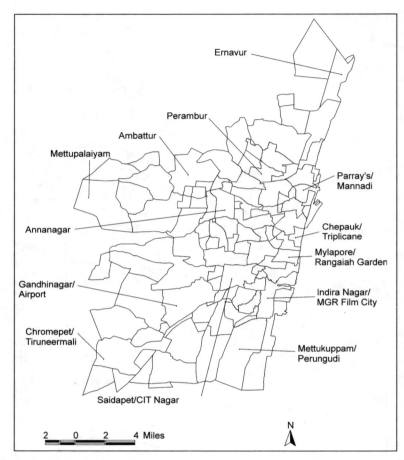

Figure 5.3: Metropolitan Chennai and its sub-regions

Source: Digital base map acquired from ML Infomap (1998).

sector. Chennai continues to be a stronger industrial centre than Bangalore. It is also part of a large industrial region in the state of Tamil Nadu along with the cities of Madurai, Coimbatore, and Salem.

ANALYSIS

We have analysed data for eight industrial sectors in three metropolitan areas for two variables (number of factories and number of workers). This means that for each basic test, 48 result points are presented. There is some danger of drowning in data because the results vary by industry sector, by metropolis, and for factories and workers. We attempt to simplify some of the data clutter in the discussion accompanying the tables and maps. It is important to note that from a theoretical perspective the most significant variable is the factories/workers distinction because it is based on the assumption that factory size has a significant bearing on location decisions. Large factories can be expected to rely more on internal economies of scale; small factories may rely on external or localization economies. Here the variable 'factories' suggests small-scale units; when factories are clustered, it suggests that a large number of small firms are clustered. Conversely, when workers are clustered, we can assume that a small number of large firms are clustered.

We present the results of the analysis in three parts. First, we test for global clustering; that is, we examine whether the distribution of factories and workers in the metropolis as a whole is clustered. Second, we test for local clustering using maps identifying local clusters and a table summarizing the map information. Third, in the longest section, we test for co-location and co-clustering of industry pairs. Each section has findings that inform both the empirical and theoretical questions identified in the introductory section.

GLOBAL CLUSTERING

The measures of global clustering (Moran's I) for the eight industry sectors for the three metropolises are reported in Table 5.1. In general, clustering is most consistent in Calcutta; Clustering in Chennai is more evident than in Mumbai, with more consistent clustering among factories than workers in Chennai. Clustering among factories and workers in the same industry and city is a common pattern; nine of the twenty-four pairs show this combination. However, combinations with only one of them (factories or workers) being clustered is equally commonly observed; in seven cases, only factories are clustered, and in two cases, only workers are clustered. This suggests, as expected, that small units are more clustered than large units. However, at this point in the analysis, the overall picture is unclear.

Table 5.1: Indices of global clustering

	Total Factories	Moran's I for factories	Z of I	Total Workers	Moran's I for workers	Z of I
Mumbai						
Food/Beverages	274	**0.172**	**2.951**	16992	**0.156**	**2.699**
Textiles	1988	**0.100**	1.797	131318	**0.191**	**3.259**
Leather	115	0.074	1.374	2629	0.018	0.474
Printing/Publishing	882	0.060	1.139	17001	0.048	0.959
Chemicals	1146	**0.149**	**2.591**	41470	0.008	0.031
Metals	894	0.069	1.290	25056	−0.019	0.145
Machinery	767	0.104	1.867	40808	−0.045	0.560
Electrical/Electronic	802	**0.244**	**4.133**	58871	**0.141**	**2.455**
Calcutta						
Food/Beverages	355	**0.165**	**3.332**	9573	**0.170**	**3.429**
Textiles	355	0.063	1.354	90621	**0.168**	**3.380**
Leather	283	**0.211**	**4.216**	6149	**0.316**	**6.229**
Printing/Publishing	428	**0.366**	**7.203**	10574	**0.255**	**5.071**
Chemicals	671	**0.214**	**4.274**	24871	**0.147**	**2.992**
Metals	1023	**0.178**	**3.582**	38440	0.047	1.067
Machinery	451	**0.130**	**2.655**	13322	**0.459**	**9.003**
Electrical/Electronic	511	**0.168**	**3.400**	18320	0.051	1.129
Chennai						
Food/Beverages	166	0.094	1.798	8459	0.069	1.353
Textiles	1446	**0.135**	**2.495**	107266	0.039	0.694
Leather	406	0.058	1.161	17919	0.087	1.666
Printing/Publishing	402	**0.230**	**4.135**	9361	**0.141**	**2.595**
Chemicals	512	**0.126**	**2.329**	17920	**0.132**	**2.443**
Metals	695	**0.184**	**3.337**	29076	0.006	0.266
Machinery	416	**0.153**	**2.799**	18772	0.039	0.829
Electrical/Electronic	359	**0.127**	**2.361**	21018	−0.003	0.098

Source: CSO Government of India, ASI sampling frame, 1998–9.
Note: Figures in bold are significant at 0.05.

LOCAL CLUSTERING

In this section, we identify the locations and other characteristics of the clusters. Recall from our earlier discussion on the measurement of clustering that local clusters may exist even when the system as a whole is not clustered (hence the distinction between 'global' and 'local' Moran indices). These results are reported in Table 5.2 and in a series of maps (Figures 5.4 through 5.6).

It is necessary to explain what is reported in these figures and the table. We began the analysis by calculating first order local Morans

Table 5.2: Concentration in clusters

	Total Factories	Factories in Clusters	Per cent of factories in Clusters	Total Workers	Workers in Clusters	Per cent of workers in Clusters	Number of Pin codes in factory clusters	Number of Pin codes in worker clusters	Common Pin codes
Mumbai									
Food/Beverages	274	79	28.8	16992	7791	45.9	5	5	0
Textiles	1988	703	35.4	131318	62649	47.7	5	6	2
Leather	115	45	39.1	2629	691	26.3	4	3	1
Printing/Publishing	882	203	23.0	17001	3567	21.0	2	2	2
Chemicals	1146	508	44.3	41470	5623	13.6	7	3	2
Metals	894	307	34.3	25056	1347	5.4	4	1	1
Machinery	767	185	24.1	40808	0	0.0	3	0	0
Electrical/Electronic	802	318	39.7	58871	34575	58.7	4	3	3
Calcutta									
Food/Beverages	355	122	34.4	9573	6034	63.0	5	9	4
Textiles	355	120	33.8	90621	38271	42.2	6	7	1
Leather	283	209	73.9	6149	4393	71.4	6	8	6
Printing/Publishing	428	209	48.8	10574	4155	39.3	10	9	9
Chemicals	671	224	33.4	24871	10053	40.4	7	9	7
Metals	1023	357	34.9	38440	8166	21.2	4	3	2
Machinery	451	101	22.4	13322	4217	31.7	6	6	1
Electrical/Electronic	511	170	33.3	18320	5283	28.8	9	6	4

Table 5.2 (contd)

Chennai

Food/Beverages	166	45	27.1	8459	3065	36.2	4	4	2
Textiles	1446	313	21.6	107266	22447	20.9	7	3	2
Leather	406	204	50.2	17919	9932	55.4	<u>4</u>	<u>6</u>	4
Printing/Publishing	402	161	40.0	9361	2534	27.1	<u>8</u>	<u>5</u>	5
Chemicals	512	137	26.8	17920	5333	29.8	**6**	**5**	2
Metals	695	240	34.5	29076	5755	19.8	4	3	2
Machinery	416	167	40.1	18772	5222	27.8	<u>2</u>	<u>2</u>	2
Electrical/Electronic	359	157	43.7	21018	1065	5.1	4	1	1

Source: CSO Government of India, ASI sampling frame, 1998-9.
Note: Pin codes with a high degree of overlap for factory and worker clusters are underlined; those with a low degree of overlap are shown in bold.

INDUSTRIAL CLUSTERS WITHIN METROPOLITAN REGIONS 159

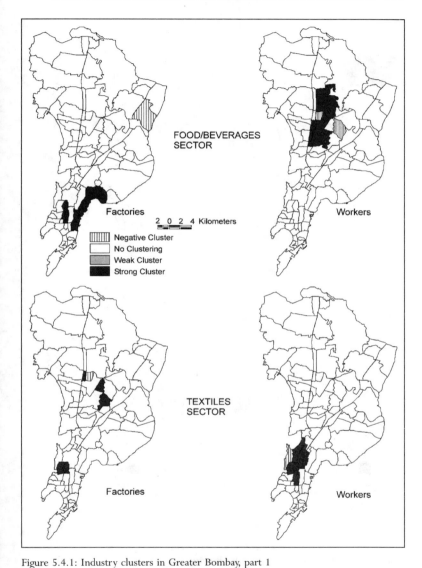

Figure 5.4.1: Industry clusters in Greater Bombay, part 1

Source: Authors' calculations from ASI sampling frame data files for 1998–9.

160 MADE IN INDIA

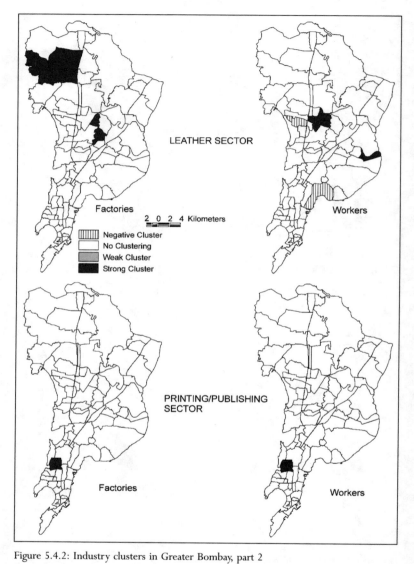

Figure 5.4.2: Industry clusters in Greater Bombay, part 2

Source: Authors' calculations from ASI sampling frame data files for 1998–9.

INDUSTRIAL CLUSTERS WITHIN METROPOLITAN REGIONS 161

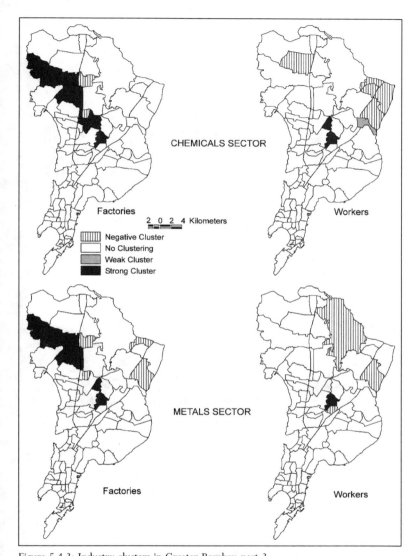

Figure 5.4.3: Industry clusters in Greater Bombay, part 3

Source: Authors' calculations from ASI sampling frame data files for 1998–9.

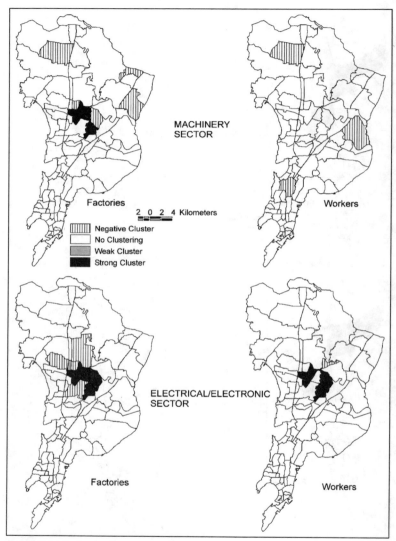

Figure 5.4.4: Industry clusters in Greater Bombay, part 4

Source: Authors' calculations from ASI sampling frame data files for 1998–9.

INDUSTRIAL CLUSTERS WITHIN METROPOLITAN REGIONS 163

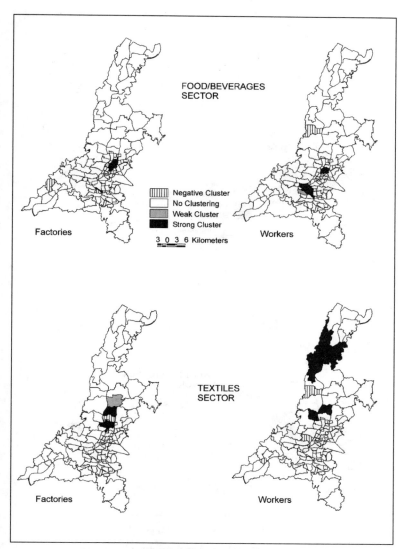

Figure 5.5.1: Industry clusters in Calcutta metropolis, part 1

Source: Authors' calculations from ASI sampling frame data files for 1998–9.

164 MADE IN INDIA

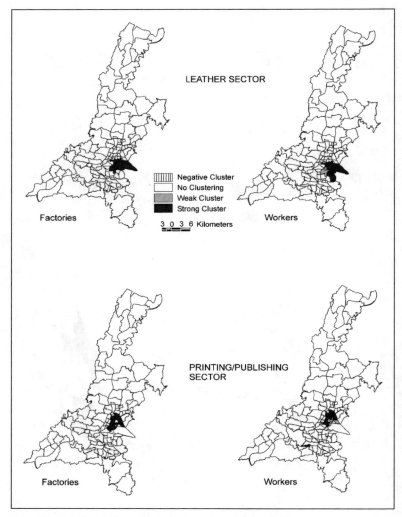

Figure 5.5.2: Industry clusters in Calcutta metropolis, part 2

Source: Authors' calculations from ASI sampling frame data files for 1998–9.

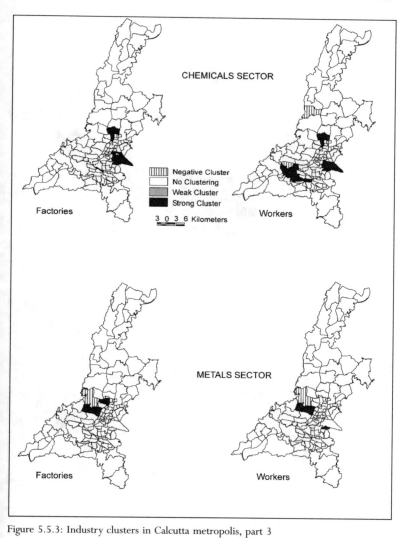

Figure 5.5.3: Industry clusters in Calcutta metropolis, part 3

Source: Authors' calculations from ASI sampling frame data files for 1998–9.

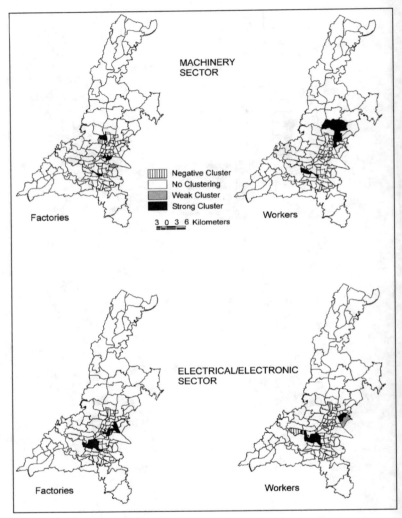

Figure 5.5.4: Industry clusters in Calcutta metropolis, part 4

Source: Authors' calculations from ASI sampling frame data files for 1998–9.

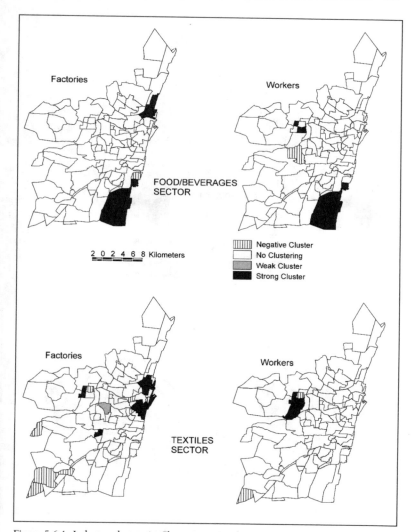

Figure 5.6.1: Industry clusters in Chennai metropolis, part 1

Source: Authors' calculations from ASI sampling frame data files for 1998–9.

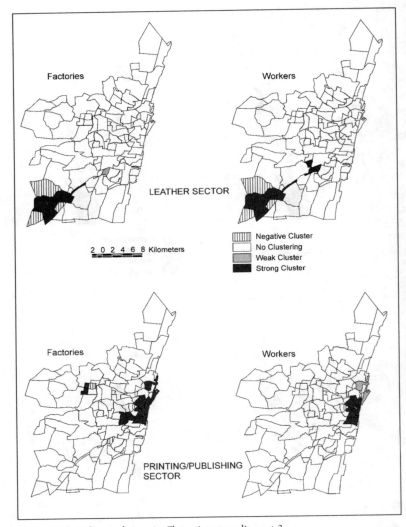

Figure 5.6.2: Industry clusters in Chennai metropolis, part 2

Source: Authors' calculations from ASI sampling frame data files for 1998–9.

INDUSTRIAL CLUSTERS WITHIN METROPOLITAN REGIONS 169

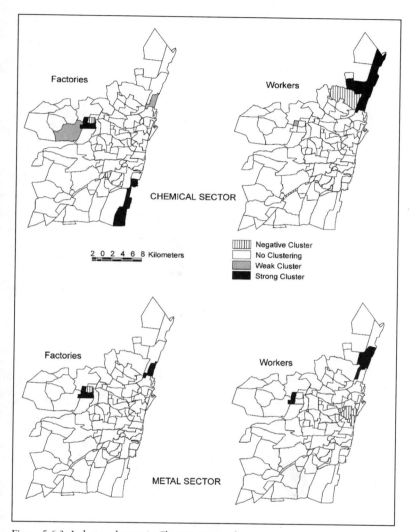

Figure 5.6.3: Industry clusters in Chennai metropolis, part 3

Source: Authors' calculations from ASI sampling frame data files for 1998–9.

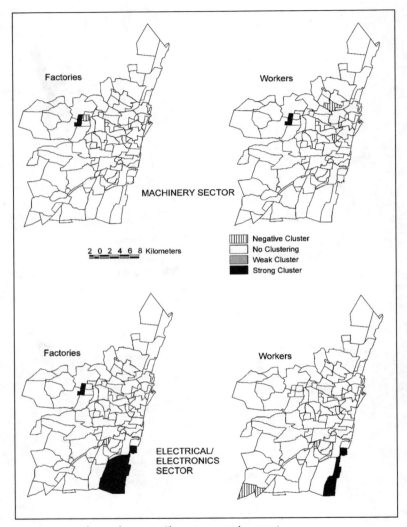

Figure 5.6.4: Industry clusters in Chennai metropolis, part 4

Source: Authors' calculations from ASI sampling frame data files for 1998–9.

for each of our eight industry categories, for each of our three cities, for factories and workers separately. In Figure 5.4.1, the first pair of figures from the top left shows the distribution of local Morans for the food and beverage sector in Mumbai; the map on the left shows the local Morans for factories, the map on the right shows the local Morans for workers. The strength of the clusters is determined by the statistical significance of each pin code's local Moran. These Z-values for local Morans are mapped. A negative cluster has Z values less than −1.65. Z-values between −1.65 and 1.65 are not clustered. Values of Z between 1.65 and 1.95 are weakly clustered, and values greater than 1.95 are strongly clustered. Pin codes that show weak and strong evidence of positive clustering are shown in the maps using different shades. In general, where local clustering does exist it does so at Z values greater than 1.95.

Finally, the numbers of pin codes forming clusters and the numbers of factories and workers in their respective clusters were totalled. These data are reported in Table 5.2. Note the great variation in the results between cities, between industries, and within cities and within industries. Let us identify some of the observed patterns.

First, factories and workers in the same industry do not necessarily cluster in the same pin codes. In the general case, there are some common pin codes and some unique pin codes for each industry. There are two types of exceptional cases: one where there is perfect overlap between factory and worker clusters (such as printing/publishing in Mumbai and machinery in Chennai); the other in situations where there are no common pin codes (such as food and beverages in Mumbai, where there are five unique clustered pin codes each for factories and for workers) or few common pin codes (such as textiles or machinery in Calcutta). This variation is seen in every city. This is an important finding. It suggests that within the same industry, small-scale operations tend to cluster together (this is where the factories are seen to cluster), often, but not always, at separate locations from large-scale operations (where workers are seen to cluster). Later, we discuss the implications of this finding.

Second, and related to the first point, it is difficult to discern whether factories are more clustered or workers in specific industries. Recall the argument that large factories rely on internal economies of scale for productivity gains, whereas smaller factories rely on external economies, at least some of which are derived from localization or clustering. Hence, in our data, we can expect factories (small scale units) to cluster more and workers (large scale units) to cluster less. If we use the percentage of factories or workers within clusters as a measure of the extent of clustering, this expectation is correct across cities in only the printing/publishing and metals sectors. The opposite is true in the food/beverages sector, and with the exception of Chennai (where the numbers are close) in the textiles sector. This may not be a significant issue as it is unclear that the measure used here is appropriate for comparing two very different types of units (factories and workers).

Third, each city has one or two industries that appear to be more clustered than others—for instance, the textiles and electrical/ electronic sectors in Mumbai, leather and food and beverages in Calcutta, and leather in Chennai. These are also the industries that generally have a significant overlap in terms of the locations of factory and workers clusters (with the exception of textiles in Mumbai). Moreover, these are also the industries for which these cities have high location quotients (LQs) at the national level.[1] The electrical/ electronic sector in Mumbai has a LQ of 3.0, the textiles sector's LQ is 1.7. The leather sector's LQ in Chennai is 4.0, in Calcutta it is 1.5. Food and beverages in Calcutta is an exception to this pattern. Another exception is the machinery sector in Mumbai; its LQ is 2.5, yet it is the only industry in any of our study cities to have absolutely no local clustering among workers (and one of the lowest levels of clustering among factories in all cities). Therefore, though this pattern cannot be generalized, there may be a causal relationship between very high levels of clustering and industry dominance at the national level. At this point, however, it is difficult to determine the direction of the causal arrow.

Fourth, the location of industry clusters generally appears to follow a couple of widely held principles: one, polluting industries are located in fringe areas, and two, such polluting industries are located in proximity. Consider the first principle. It seems obvious that any local regulatory agency will direct the location of polluting industry toward the urban fringe. The two most polluting sectors considered here are chemicals and leather. In all three cities, these industries are located in the fringe areas, and it appears from the maps that in the cases of

Mumbai and Calcutta, these two industries share common locations. Also note the particularly interesting case of the chemicals sector in Chennai, where factories (which implies small scale units) are clustered in the southern extremity and in the far west, whereas workers (or large scale units) are located in the northern fringes of the metropolitan area.

CO-CLUSTERING, CO-LOCATION, AND INDUSTRIAL DISTRICTS

In this section, we test for the existence of co-location or co-clustering of different industry sectors. A brief discussion of theory, on why co-location or co-clustering may be possible, is necessary before we discuss the tests and outcomes. It is useful at this point to distinguish between co-location and co-clustering. Co-location occurs when industries from two sectors are present in the same neighbourhood. Co-clustering occurs if both industries that are co-located are related through economic (input–output, innovation, or labour market) linkages.

THEORY

The idea of the industrial district goes back to Marshall (1919) who suggested that small specialized firms would tend to cluster in space to derive external economies to offset the internal scale economies of large factories. Piore and Sabel (1984) argued that the late twentieth century had seen the arrival of a 'second industrial divide' where the vertically integrated organization of production characteristic of Fordist manufacturing was giving way to regional specialization organized around networks of small scale producers. Geographers see this in terms of the need for flexible specialization in globalized production systems geared toward rapid changes in technology and the need to respond to shifting patterns of demand (Amin 2002). Economists have focused on the specific productivity advantages provided by proximity. In the simplest terms, these localization economies (to be distinguished, as discussed earlier, from urbanization economies that accrue to all firms in an urban area) arise from two sources: local labour markets and knowledge spillovers.

Labour markets: Do thick local labour markets create localization economies? There are two questions inherent in this issue: Are labour markets spatially segregated by skill or specialization? Do industries become spatially segregated in response to segregated labour markets? We need to distinguish between industries based on unskilled labour (e.g., leather) from those based on skilled labour (printing/publishing). Consider labour markets with unlimited supplies of unskilled labour operating in land markets where land rents decline with proximity to industry. In other words, unskilled labour is available in plenty and can locate anywhere, but is likely to locate near factories where rents are lowest. Therefore, unskilled labour is likely to co-locate with industrial clusters. Does this generate localization economies for those industries that are reliant on unskilled labour? If it does, we are likely to see industries based on unskilled labour to be more oriented toward locating in industrial clusters with other industries based on unskilled labour that do not necessarily share any other characteristics. These clusters will not include industries based on skilled labour. If these latter (skill-based) industries do cluster, they do not do so for labour market localization economies, as their critical workers, the high-skill high wage labour, will not co-locate with industry. They do so for other reasons (discussed below). In summary, labour market localization economies, if any exist, are likely to be industry specific and inversely related to the proportion of skilled labour in a given industry.

Knowledge spillovers: These are of two kinds—technology spillovers through informal interaction and information spillovers on market agents. Technology spillovers are irrelevant in low technology firms and industries in mature stages of the product cycle. The vast majority of manufacturing industry in India belongs in this category. Hence, for these industries we can argue that there are negligible localization economies from technology spillovers. Information spillovers on market agents such buyers and suppliers, however, are more likely to provide localization economies. Firms of all sizes (except perhaps the very largest vertically integrated firms) rely on dense buyer–supplier networks. Firms benefit from having access to local buyers and suppliers, and knowledge pooling on buyer–supplier behaviour is likely to eliminate inefficient agents.

Hence, theory suggests the existence of two types of industrial districts: labour sharing industrial districts that depend on the local availability of low skill labour, and buyer–supplier linked industrial districts where industries that have market interactions with each other benefit from co-location. We do not have spatially disaggregated wage data to test explicitly for the existence of labour sharing districts. We are, however, able to establish what the expected buyer–supplier links are between our industry groups by assuming that the input–output links at the national level are replicated at the local level. The national input–output tables are available. We assume that similar input–output linkages exist at the level of the metropolis and argue that industries with strong input–output linkages are likely to co-locate.

EVIDENCE

Before we examine the evidence on buyer–supplier linkages, let us first examine the data on co-location by industry for all sectors. The correlation coefficients for factories and workers for all eight industry groups are reported in Table 5.3. The numbers here are quite remarkable. There is strong evidence that industry groups co-locate at the pin code level, especially in the case of factories, or small scale units. In Mumbai, for instance, factories for every industry group are seen to have a statistically significant correlation with every other industry group. In general, the correlation coefficients are very high: 0.93 between machinery and chemicals, 0.91 between metals and chemicals, etc. In the case of workers, however, the correlations are not as high, and fewer are statistically significant. In Chennai, the pattern is even more pronounced. With the exception of the leather sector (which has a moderate but significant correlation with only the textiles sector), the correlation coefficients of every other pair of industries is significant, and as in the case of Mumbai, often very high. The leather sector is anomalous in Calcutta as well. Its only significant correlation in that metropolis is with the chemicals sector. But Calcutta itself is somewhat of an anomaly compared to Mumbai and Chennai, at least in terms of the co-location of workers. Factories in Calcutta, with the exception of the leather sector, are generally co-located; however, workers in at least four industries—food/beverages,

Table 5.3: Correlation coefficients for industry pairs

	Food/ Beverages	Textiles	Leather	Printing/ Publishing	Chemicals	Metals	Machinery	Electrical/ Electronic
Mumbai								
Food/ Beverages	<u>1.00</u>	**0.39**	**0.31**	**0.39**	**0.34**	**0.40**	**0.40**	**0.25**
Textiles	0.10	<u>1.00</u>	**0.58**	**0.79**	**0.77**	**0.73**	**0.77**	**0.63**
Leather	**0.20**	0.23	<u>1.00</u>	**0.43**	**0.77**	**0.63**	**0.75**	**0.63**
Printing/ Publishing	0.16	**0.74**	**0.35**	<u>1.00</u>	**0.49**	**0.44**	**0.51**	**0.45**
Chemicals	**0.25**	**0.37**	**0.42**	**0.33**	<u>1.00</u>	**0.91**	**0.93**	**0.81**
Metals	0.16	**0.26**	0.18	**0.25**	**0.56**	<u>1.00</u>	**0.87**	**0.70**
Machinery	0.09	0.17	0.15	0.07	**0.34**	0.15	<u>1.00</u>	**0.78**
Electrical/ Electronic	**0.27**	0.07	0.16	0.10	**0.21**	**0.25**	0.06	<u>1.00</u>
Calcutta								
Food/ Beverages	<u>1.00</u>	**0.31**	0.09	**0.46**	**0.43**	**0.33**	**0.46**	**0.40**
Textiles	0.07	<u>1.00</u>	0.04	**0.36**	**0.64**	**0.33**	**0.51**	**0.51**
Leather	0.02	−0.05	<u>1.00</u>	0.03	**0.49**	0.07	0.12	0.10
Printing/ Publishing	0.12	0.03	0.08	<u>1.00</u>	**0.34**	0.15	**0.37**	**0.32**
Chemicals	0.17	**0.19**	**0.33**	0.10	<u>1.00</u>	**0.36**	**0.60**	**0.72**
Metals	0.14	0.09	0.14	0.07	**0.25**	<u>1.00</u>	**0.76**	**0.33**
Machinery	**0.21**	−0.03	0.04	**0.26**	**0.22**	**0.41**	<u>1.00</u>	**0.68**
Electrical/ Electronic	**0.25**	0.06	0.02	0.08	**0.37**	0.08	**0.30**	<u>1.00</u>
Chennai								
Food/ Beverages	<u>1.00</u>	**0.30**	0.02	**0.47**	**0.69**	**0.59**	**0.52**	**0.62**
Textiles	0.15	<u>1.00</u>	**0.29**	**0.61**	**0.51**	**0.40**	**0.41**	**0.40**
Leather	−0.01	**0.35**	<u>1.00</u>	0.10	0.16	0.11	0.14	0.15
Printing/ Publishing	**0.25**	**0.32**	0.17	<u>1.00</u>	**0.58**	**0.56**	**0.57**	**0.46**
Chemicals	**0.27**	**0.34**	**0.23**	**0.33**	<u>1.00</u>	**0.87**	**0.84**	**0.79**
Metals	**0.22**	**0.31**	0.13	**0.63**	**0.33**	<u>1.00</u>	**0.95**	**0.71**
Machinery	**0.41**	**0.39**	0.09	**0.31**	**0.41**	**0.34**	<u>1.00</u>	**0.76**
Electrical/ Electronic	**0.23**	**0.50**	**0.27**	**0.26**	**0.40**	**0.23**	**0.27**	<u>1.00</u>

Source: CSO Government of India, ASI sampling frame, 1998–9.
Notes: Factory data above the diagonal, Worker data below the diagonal.
Figures in bold are significant at 0.01.

Table 5.4: Indices of co-clustering in selected industry pairs

	Factories				Workers		
	National IO factor	Total Factories	Moran's I	Z of I	Total Workers	Moran's I	Z of I
Mumbai							
Strong IO links							
Metals and Machinery	34.95	1661	0.082	1.496	65864	−0.052	0.668
Metals and Electrical/ Electronic	25.75	1696	**0.161**	**2.774**	83927	0.105	**1.876**
Weak IO links							
Food/Beverages and Electrical/Electronic	0.07	1076	**0.184**	**3.145**	75863	**0.294**	**4.936**
Textiles and Metals	0.41	2882	0.080	1.473	156374	0.045	0.904
Calcutta							
Strong IO links							
Metals and Machinery	34.95	1474	**0.222**	**4.429**	51762	**0.121**	**2.49**
Metals and Electrical/ Electronic	25.75	1534	**0.211**	**4.222**	56760	**0.109**	**2.249**
Weak IO links							
Food/Beverages and Electrical/Electronic	0.07	866	**0.178**	**3.579**	27893	**0.185**	**3.730**
Textiles and Metals	0.41	1378	**0.270**	**5.349**	129061	**0.181**	**3.644**
Chennai							
Strong IO links							
Metals and Machinery	34.95	1111	−0.079	1.060	47848	0.029	0.673
Metals and Electrical/ Electronic	25.75	1054	**0.149**	**2.738**	50094	−0.004	0.087
Weak IO links							
Food/Beverages and Electrical/Electronic	0.07	525	**0.126**	**2.332**	29477	0.041	0.873
Textiles and Metals	0.41	2141	**0.161**	**2.948**	136342	0.026	0.623

Source: CSO Government of India, ASI sampling frame, 1998–9.
Note: Figures in bold are significant at 0.05.

textiles, leather, and printing/publishing—are generally not co-located with workers in other industries. These findings provide further evidence that small units are more likely to cluster than large ones.

It is one thing for industries to be co-located, it is another for them to be co-clustered. If the reasoning on buyer–supplier clusters outlined above is correct we can expect: (1) that industries that have strong input–output links will co-cluster; that is, they will form clusters in the same or proximate pin codes, and (2) industries that do not share buyer–supplier linkages will not co-cluster; if they do it will be for labour sharing, and therefore they will share similar labour profiles (low-skill with low-skill, or high-skill with high-skill). In order to test this hypothesis, we conducted co-clustering tests on four industry pairs; two of these pairs have strong input–output linkages (the two highest among our eight industry groups); metals and machinery with an IO factor of 34.95, and metals and electrical/electronic with an IO factor of 25.75. The two other pairs have virtually no input–output linkages; food/beverages and electrical/electronic have an IO factor of 0.07 and very different labour profiles (the former is low skill, the latter is high-skill); textiles and metals have an IO factor of 0.41 and potentially similar labour profiles. We combine the proportion (rather than raw) data on factories and workers for these four industry pairs by pairs, and conduct tests of global clustering (Moran's I) and local clustering.

The results are mixed. Consider the results of the global clustering test first (in Table 5.4). In Mumbai, among the strong IO pairs, only factories in the metals–electrical combination are clustered, whereas the food–electrical combination, which shares neither input–output IO links nor similar labour profiles, shows clustering for factories and workers. In Calcutta, all four pairs are clustered, for factories and workers (more for the former than the latter). In Chennai, as in Mumbai, factories in the metals–electrical combination are clustered; but factories in both weak IO combinations are clustered. The strongest IO pair—metals and machinery—is not only not clustered in Mumbai and Chennai, but the value of Moran's I is negative half the time.

Next, for further evidence on co-location, we look at a final set of data: the share of industry in the top pin codes. Table 5.5 lists these data by metropolis, for all industry and by sector. (Note that in a given metropolis the top 10 or top 20 pin codes are not separately identified for each sector. What is reported is the share of each sector in the top overall pin codes.) The results are not surprising in the context of what we have presented earlier, but they are quite effective in making the point that industry is concentrated in a handful of pin codes in all three cities, and that most sectors are heavily represented in these top pin codes. In Mumbai, the top ten pin codes include close to 55 per cent of all factories and workers, and the top 20 pin

Table 5.5: Industry concentration in top districts

	Number of units	Share of metropolitan total (%)	Food/ Beverages (%)	Textiles (%)	Leather (%)	Printing/ Publishing (%)	Chemicals (%)	Metals (%)	Machinery (%)	Electrical/ Electronics (%)
Mumbai Factories										
Top 10	3744	55.20	23.73	57.75	56.52	45.69	57.77	60.40	56.32	53.49
Next 10	1447	21.34	25.18	22.69	19.13	16.78	21.73	20.69	23.99	17.33
Top 20 Total	5191	76.54	48.91	80.43	75.65	62.47	79.49	81.10	80.31	70.82
Mumbai Workers										
Top 10	176701	53.19	28.05	51.18	20.27	35.45	34.17	50.00	62.75	77.91
Next 10	86411	26.01	15.70	32.58	51.16	23.27	29.80	24.15	25.16	11.85
Top 20 Total	263112	79.21	43.75	83.76	71.44	58.72	63.97	74.15	87.91	89.76
Calcutta Factories										
Top 10	1610	36.38	24.35	38.64	59.25	26.44	29.70	46.22	35.54	26.37
Next 10	787	17.79	25.13	13.05	21.57	16.78	18.08	13.72	18.88	22.41
Top 20 Total	2397	54.17	49.48	51.70	80.82	43.22	47.78	59.93	54.42	48.78
Calcutta Workers										
Top 10	77588	33.73	11.19	57.66	0.42	7.00	19.81	19.02	9.68	25.57
Next 10	49117	21.35	30.47	23.14	42.00	15.09	13.98	23.82	13.33	13.40
Top 20 Total	126705	55.09	41.66	80.80	42.42	22.09	33.79	42.84	23.01	38.97
Chennai Factories										
Top 10	1872	42.53	33.74	27.25	53.20	28.36	40.62	51.22	69.95	66.02
Next 10	792	17.99	21.08	23.03	6.16	33.83	18.16	17.12	5.53	7.80
Top 20 Total	2664	60.52	54.82	50.28	59.36	62.19	58.79	68.35	75.48	73.82
Chennai Workers										
Top 10	104902	45.65	23.34	40.60	53.58	35.54	42.91	51.12	44.82	73.68
Next 10	45364	19.74	29.32	20.44	14.09	14.56	15.17	16.37	38.04	11.66
Top 20 Total	150266	65.39	52.66	61.04	67.66	50.10	58.08	67.49	82.85	85.34

Source: CSO Government of India, ASI sampling frame, 1998–9.

codes include over 76 per cent of all factories and about 80 per cent of all workers. In Chennai and Calcutta, the proportions are progressively smaller. This decline is probably a function of the fact that the total number of pin codes in Calcutta (133) and Chennai (108) are higher than in Mumbai (94). In general, regardless of the number of pin codes in a metropolis, the top 10 per cent of the pin codes include close to 50 per cent of all factories and workers.

SUMMARY OF THE FINDINGS

(1) In Indian metropolises, industry is generally clustered—the evidence for clustering is found at the level of the metropolis (using the global Moran statistic), and at the level of the pin code (using local Moran statistics and maps). At the sectoral level, there are extremes: in one case there are no clusters at all; at the other end there are instances where 70 per cent or more of workers/factories are concentrated within six to eight pin codes (see Table 5.2).

(2) The clusters are of two types: one, where factories and workers are both clustered in the same region within a metropolis; two, where factories are clustered at locations separate from worker clusters (that is separate clusters of small scale operations and large scale operations). Both patterns are equally common. In general, factories (small units) are more clustered than workers (large units).

(3) A small number of pin codes account for a very large proportion of all industry, both factories and workers. As a result, the extent of industry co-location is very high. However, the expected relationships between industrial sectors—whereby industries with strong input–output linkages are expected to co-locate, and industries using similar labour profiles are expected to co-locate—are not found. On the contrary, we see several examples of counter-intuitive co-locations.

(4) Some industries have distinct locational properties. For example, where the leather industry is significant (as in Calcutta and Chennai), it is located on the urban fringe and is not co-located with other industries except chemicals. The printing/publishing industry is located near the urban core in all three cities. Both

industries are also highly clustered. The textile industry, on the other hand, the largest industry, is marked by separate factory and worker clusters, where the former is closer to the city centre than the latter (except in Mumbai).

AN EXPLANATORY FRAMEWORK

How can these patterns be explained? To begin with, we suggest that the conventional wisdom—according to which localization economies drive cluster formation—may be limited in its explanatory power. There is little evidence in support of the processes of localization, either via local labour markets or via local buyer–supplier networks. It is possible to argue that these findings are artifacts of the method of local cluster identification. That is, had we used other methods that would have enlarged the definition of 'local' beyond the one used here (a pin code and its adjacent neighbours), there would have been stronger evidence in support of localization economies driving cluster formation. That, however, is unlikely. First, at a preliminary stage in our investigations we did indeed use larger definitions of 'local', with very similar results. Second, it may be possible to enlarge the definition of local till it is meaningless or large enough to encompass a significant portion of the metropolitan area. At that point, it becomes difficult to distinguish localization economies from urbanization economies.

One can also argue that localization economies arise primarily as a result of inter-firm trade or technological interaction within an industrial sector as defined in this analysis. That is, the sectors have been so broadly aggregated that most trade and technical exchange takes place within rather than between sectors. For many sectors, it is difficult to make that case, since according to the national input–output data, inter-firm trade within most of these sectors is not very high. For instance, within the food/beverages, chemicals, and machinery sectors the intra-sector input–output coefficients are under 5.5. The other sectors, where intra-sector input–output links are stronger, are also dominated by old firms that are not close to the technological frontier (Lall and Rodrigo 2001); as a result, there is little likelihood of intra-sector technical exchange. Therefore, we must conclude that buyer–supplier networks and labour pools are metropolis wide rather than localized; hence, the debate on the relative strength of localization and urbanization economies for industrial location decisions should be resolved firmly in favour of the latter.

Looking beyond localization economies, we suggest that firm locations are guided by a complex set of factors which often rule out most spaces within metropolitan areas. These factors include the accidents of history, metropolitan expansion, industry characteristics, and, above all, state regulations (especially ones that affect the land market). As a result, we see the unplanned or planned evolution of mixed industrial districts, which include a variety of related and unrelated industry sectors, at leapfrogging locations within metropolitan areas. We suggest that a historical framework may explain the observed patterns of industry location in Indian metropolises. This framework is speculative (we do not have the temporal information needed to make a stronger claim), but it is able to account for many of the expected and unexpected findings.

First, at some specific historic point, the pioneer industrial unit in a specific industrial sector makes a location decision within the metropolis. The driving force behind that decision (not the location decision, but the decision to start a new industry) may have one or several motivations: proto-nationalism (as in the case of the first textile mill in 1854 in Mumbai), war (as in the case of the first leather units in Calcutta, created to outfit saddles for the Imperial army during World War I), bureaucratization and the spread of literacy (creating the need for printed matter), etc. This first factory did not rely on localization economies to boost productivity, but probably relied on urbanization economies, at least in terms of providing market access and a pool of labour. This unit located in what was then the urban fringe, beyond dense population settlements, but close enough for workers to reach the plant. Following the convention of the time, this unit was large, relying on internal economies of scale to reduce costs.

A cluster of firms in the same industry began forming around this original location. It is difficult to determine whether these subsequent location decisions were the result of localization advantages arising from labour pooling, advantages derived from shared infrastructure such as railheads, or state regulations that directed new firms to this location. It will be necessary to conduct archival research for more concrete judgments. It is very likely that at some point industries in this cluster began to derive benefits from labour pooling. Two factors should be considered here. First, these were not the high technology industries of the time; they were industries in the late stages of the product cycle (Vernon 1966), reliant on unskilled and semi-skilled labour. Second, the location of these industries began influencing the land market around them; because of the local environmental impact of these industries, the primary bidders for the proximate land were other industries or low wage labour. In other words, this sub-region became an industrial district, with large-scale factories and slums for unskilled workers. As the city continued to grow beyond this industrial district, most new industry was directed to this district by state regulations. The single purpose industrial district became a mixed industrial district.

Where was even newer industry to locate? With all the land in the original industrial district in use, and with state regulations that forbade the conversion of any land with housing (slums, tenements, or middle class residences) to industrial use, new industries now sought new locations on what was then the urban fringe. The role of the state in discouraging land use change turns out to be a critical influence on industrial location (as we shall see below). The cycle of industrial district formation began again, this time with more active involvement of the post-colonial state which assisted with land acquisition (often the most difficult aspect of industrial location in urban areas) and provided physical infrastructure.

When this cycle was complete (that is, the point was reached when there was no more land for new factories), new industrial units leapfrogged over the residential communities that had grown in the interim to new locations at the urban fringe. This is the current stage. Now there is even more active involvement of the state, which sets up export processing zones, free trade zones, technology parks, industrial parks, etc. to entice new industrial units. (Consider the names of some of the most active pin codes. In Mumbai: Chakala and SEEPZ [Santa Cruz Electronics Export Processing Zone], both Maharashtra Industrial Development Corporation centres; in Chennai: Ambattur Industrial Estate and SIDCO Industrial Estate, both set up by the state of Tamil Nadu.)

This stylized framework explains a number of observations—some regularities, some irregularities—listed in the preceding section. In this framework, the exact location of specific industries has to be understood in terms of the functions of these industries. For instance, the printing/publishing industry remains located close to the CBD (Central Business District), which is its principal market area. The leather industry, which pollutes both the air (with the smell of animal hides) and the groundwater, and therefore cannot even co-locate with other industries (not to mention residential areas), remains at the urban fringe—when the fringe moves beyond or envelopes the leather cluster, the state compels the cluster to move further out. This is exactly what has happened in Chennai and is happening in Calcutta.

This framework also explains some anomalies, such as the location of a large textile cluster near the Mumbai CBD. This cluster should not exist, for the land there is too valuable to remain devoted to an industry with outmoded technology and very low value addition. Yet it remains a textile cluster, with virtually closed factories, state takeover (or nationalization) of 'sick' units, and job losses that mount by the year, because the state will not permit the conversion of this industrial land to commercial or residential use (D'Monte 2000).[2]

The other significant anomaly comes from the relative locations of factory and worker clusters. It is not unusual to see these clusters form separately, but the expectation is that worker clusters (which are clusters of large factories), which require more undivided land, will locate on the least expensive land, furthest from the centre. Yet when there is no exit policy, that is, factories are not allowed to close and factory land cannot easily be transferred, it is possible to see clusters of large scale units near the centre of the city. (Recall that many of the early industrial units, following the Fordist principles of the times, were large units.) And most important, this framework explains why co-location is so common, but theoretical expectations on co-clustering are not realized in practice.

CONCLUSION

Our principal conclusion is that land market rigidities created by state policy rather than market opportunities in the form of localization economies have paramount importance in guiding intra-metropolitan industrial locations. The key variable is land-use policy: segregationist/environmental policies that isolate polluting industries like leather and chemicals; the absence of exit policies and land-use change policies that keep open obsolete large factories close to the

CBD; activist industry attraction/promotion policies that lead to the creation of new industrial zones which include a variety of industry sectors. This can be a serious problem in developing location theory which assumes the operation of unhindered market forces, especially in the land market. The evidence suggests that intra-metropolitan industrial location decisions are significantly influenced by state regulations. Does this mean that firms cannot utilize spatial economies? Do state interventions neutralize and perhaps even negate the spatial economies that would be possible in unfettered land markets? In theory, this possibility certainly exists. However, the fact is that industries do end up being clustered. In fact, by reducing the location choices for firms, state policies probably help in cluster formation. If spatial economies in production arise from being clustered, irrespective of whether the clustering came about as a result of market forces or state action, then we must conclude that these firms can enjoy at least some spatial economies. That is, state interventions influence industry location in significant ways, but, having led to industrial clustering, should have relatively little impact on the generation of external economies during the production phase.

We must remember (from Chapter 3) that there is a significant disconnect between the 'revealed preference' of firms (expressed in clustering) and their 'declared preference' in survey-based studies that show that there is a substantial random element in the choice of location: personal reasons, chance, and opportunity (especially that of finding a good site) are given as explanations almost half the time, proximity to other similar firms is not a serious factor (see Calzonetti and Walker 1991; Mueller and Morgan 1962). Therefore, it may be useful to reconsider some aspects of location theory from the perspective of what institutional economists call habit (Hodgson 1998) and social theorists call habitus (Bourdieu 2002). These theories suggest that most decisions are made on the basis of perceptions and are characterized by imitation, inertia, and cumulative causation. These are, nevertheless, efficient decisions because imitation and cumulative causation reduce risk and the cost of decision-making. When land markets are constrained, it makes sense to locate where other firms locate. The benefits of proximity may follow during the production phase, but during the location decision phase, firms may be more interested in risk reduction than in seeking localization economies.

NOTES

1. The LQ is simple measure of regional concentration used in regional science. It calculates the ratio of the share of a given variable to the share of population. Here, LQ = 1 indicates that a pin code's share of a particular sector is equal to its share of all industry. If LQ = 3, it indicates that the pin code's share of that sector is three times its share of all industry.
2. As we write, this pattern appears to be finally breaking. Textile mill land in the heart of commercial Mumbai is changing hands, being sold for astronomical sums to real estate development companies.

6

On Spatial Policy

So, what can Bihar do? Even the most casual observer of the Indian situation knows that development outcomes in Bihar significantly lag behind the rest of the country. Bihar has the highest poverty incidence, with 40 per cent of its population in 1999 being under the poverty line (World Bank 2004a). Net primary school enrollment ratio in 1999 was 52 per cent (compared to the Indian average of 77 per cent); the female literacy rate of 33.6 per cent was far below the already low Indian average of 54.3 per cent; and 10.3 per cent of households had access to electricity as a source of lighting (the Indian average was 55.8 per cent). In Chapter 2 we have shown that the average income in Bihar is about half the Indian average, which makes it close to or less than one-third of the averages of several leading states. Industrial capital formation in Bihar has gone down from almost 13 per cent of the national total to about 3.7 per cent in the post-reform period. Much of this was in a single district, Purbi Singbhum, which is no longer part of the state (having become part of the new state of Jharkhand). In short, Bihar, for long the poorest state in the union, is falling even further behind, perhaps at an even more rapid rate.

This condition is not unique to India or Bihar. For example, in Brazil, per capita income differences across regions are relatively large and surprisingly stable over very long periods. Per capita income in the prosperous Southeast was 2.9 times that of the lagging Northeast in 1939 and 2.8 times in 1992 (Lall and Shalizi 2003). At a finer spatial scale, regional differences in per capita income are much more pronounced, with per capita incomes in São Paulo (the wealthiest southeastern state) being 7.2 times that of Piaui (the poorest northeastern state). Further, of the ten poorest states in the country, eight are in the Northeast, and two in the North region (Azzoni *et al.* 2000). In Indonesia, in the mid-1990s, the ratio of average income between Jakarta (the richest province) and Central Java (the poorest province within the Java region) varied between 4.9 and 5.4. In the country as a whole, the ratio of income in Jakarta to the poorest province (East Nusa Tenggarra, in the outer islands) varied between 8.3 and 9.5. In Mexico, the ratio of per capita GDP between the Federal District (where Mexico City is located) and Oaxaca (in the poor southern region) went from 5.4 in 1970, to 3.7 in 1985, to 5.7 in 1999. The same ratio with the state of Chiapas (where the revolutionary Zapatistas operate), a better known case of regional inequality and consequent insurgency, went from 3.9 in 1970, to 2.2 in 1980, 2.7 in 1985 and 5.8 in 1999. (These data are all reported in Chakravorty 2005.)

Therefore, when we ask the question 'what can Bihar do?' we have to look for answers not merely in the Indian context, but in the context of what other nations have attempted to do in order to mitigate regional imbalances. We focus on industrial policy, of course, because our entire effort up to this point has concentrated on industrialization as the primary source for raising productivity and income. In this chapter, we take a critical look at what policy instruments are available, what have been used in India, and what can succeed, to what extent. The conclusions from this critical analysis are not very hopeful, for we see little that can be done to improve conditions in lagging regions, dramatically, in short time periods. The burdens of history and geography are heavy indeed. But before we do that, let us begin with a brief summary of what we have found so far:

(1) The pattern of industry location went through one well-known major shift in the 1960s when the leading state (West Bengal) began to decline precipitously. This resulted in what appeared to be a decline in regional inequality up to the late 1970s.

(2) From the 1980s, the overall trend has remained unchanged. We have called this a pattern of concentrated decentralization, or inter-regional divergence with intra-regional convergence. Industries have continued to locate within larger and more expanded leading industrial regions (with the exception of Calcutta) so that there is deconcentration within these leading

regions at the same time that there is increasing inter-regional concentration.
(3) These trends have accelerated after the liberalizing structural reforms. One of the important reasons for this is the decline of the state industrial sector and the increasing dominance of the private sector. The latter has a clear bias toward coastal regions and existing industrial regions.
(4) Private firms derive definite cost advantages from locating within regions that have a diverse industrial base. The cost-reducing effects of this local industrial diversity is most pronounced for small firms. Localization economies (created by the presence of firms in the same industry) appear to be virtually non-existent.
(5) Within metropolitan regions too, localization economies do not appear to have a strong influence on the location of factories. Therefore, general urbanization economies, or industrial diversity can be presumed to be the stronger centripetal force. The most important factor is the local land market and state actions to influence that market.

THE POLICY ENVIRONMENT

Many governments have opted for directed interventions to offset some of the market forces that lead to clustering and uneven development in order to promote relatively balanced regional development. This has created the policy tension between the market solution of out-migration or labour flows—'moving people to jobs'—and the interventionist solution of 'moving jobs to people' by promoting capital flows as well as providing infrastructure and other public goods. Over the past 50 years, federal and sub-national governments in many countries have designed policies and programs aimed at reducing disparities in economic performance among their regions. These include developed countries such as Japan, the United States and the United Kingdom, and developing countries such as Brazil, Indonesia, Malaysia, Mexico, and Thailand. While some public interventions have been topic- or sector-specific national programmes which tend to have spatially differentiated effects due to initial conditions and lags, much of the focus of regional development activities has been on spatially targeted programmes.

The first objective in this chapter is to examine the effectiveness of a limited set of instruments that have been used to support the policy objective of 'moving jobs to people'. These include public expenditures in infrastructure to improve market accessibility of lagging regions, and promoting capital flows, including fiscal transfers designed to support incomes or to subsidize the creation of jobs and the extension of credit in poor areas. There are two main questions that we want to examine in this context:

(1) Have regionally focused interventions succeeded in improving employment prospects (thereby incomes) in lagging regions?
(2) What are the implications of these interventions for national economic performance and efficiency—in particular, is there a trade-off between promoting regional equity and national economic efficiency?

While both questions are worth examining, it is difficult to say which question is the more important one. From the perspective of lagging regions, it is important to examine the performance of regionally directed interventions in terms of their stated objectives, that is, have these programmes stimulated economic performance in lagging areas? From a national efficiency perspective, the question on welfare effects of regional interventions is definitely central for designing appropriate federal policy interventions. Thus, it becomes essential to move from the limited question of 'do targeted interventions affect industry location decisions' to the more policy-relevant question of 'are these interventions welfare enhancing'. In answering the latter question, it becomes important to determine (1) whether regional interventions induce new economic activity into lagging regions or simply re-allocate businesses away from other regions, and (2) what is the effect on the productivity of re-located industries.

The empirical evidence for developing countries on either set of questions is limited, and the results are mixed. While there is an abundant and growing body of literature that examines convergence of aggregate economic and social indicators across regions, the findings from this body of research do not help us in understanding why economic activity is spatially concentrated, or the role that policy instruments can play in offsetting spatial concentration and stimulating industrial growth in lagging regions.

Convergence type analyses are typically based on a 'neoclassical' view of the world, which suggests that underlying economic forces will tend to generate relatively smooth convergence of per capita incomes as capital and technology flow to low cost locations. Even though poor endowments of skills and institutions place some locations at a disadvantage, the prediction tends to be that there will be convergence. Locations are assumed to grow 'in parallel', with faster growing lagging regions eventually catching up to dynamic ones. In contrast, the presence of clustering and agglomeration forces give rise to cumulative causation, suggesting that growth may be 'in sequence' rather than in parallel. This leads to increasing regional disparities, as some cities and regions benefit from cumulative causation processes while others are left out. In this context, it becomes important to understand the factors that influence location decisions of industry—preferably at the level of individual firms, and then examine if and how policy instruments can compensate for the 'disamenities' of locating in lagging regions, thereby improving their investment attractiveness.

We look at a narrow but commonly used set of regional policy instruments: regional incentives and infrastructure investments. In this review, we try to examine how scale economies from agglomeration and market access counteract the effectiveness of regional policies designed to stimulate economic activity in lagging regions. In addition, we argue that historical institutions and initial natural advantage lead to the formation of current agglomerations. And finally, we discuss options for improving the economic fortunes of spatially and historically disadvantaged regions.

The rest of this chapter is organized as follows. In the next section, we examine the experience from several developing countries in terms of policies to stimulate economic activity in lagging regions. We develop a typology of interventions, and then review the relative performance of these programmes. In the following section, we examine why regional policies may provide a limited 'bang for the buck'. We argue that natural agglomeration tendencies counteract policies aimed at industrial dispersion, that those policies are often contradictory, so that coordinated regional programmes which offer a package of interventions are likely to do better than individual programmes or policies; we also note the problem of institutional path dependence. In the final section, we take the main findings from the review and apply it to a central lagging region. That is, we return to the question posed at the beginning of this chapter: what are the economic prospects for a spatially and historically disadvantaged region such as Bihar?

POLICIES FOR REDUCING REGIONAL DISPARITIES

Federal and sub-national governments in many developing countries have designed and implemented spatially explicit policies to improve the economic performance of their lagging regions. Public policies in this context have typically included public expenditures in physical infrastructure to develop inter-regional linkages as well as local comparative advantage,[1] fiscal incentives to attract firms with multiplier effects, or a combination of the two. An explicit motivation behind large scale infrastructure investments is the view that infrastructure is an intermediate public good with an active role in the production process. Thus, increasing the stock of infrastructure in lagging regions, like increasing any other stock of capital, will improve the productivity of existing firms and attract new firms, thereby helping these regions grow closer to more developed ones (Puga 2002). Examples of infrastructure-led development at the sub-national level include the development of secondary cities in Malaysia and Thailand, transportation capacity development in the lagging Brazilian Northeast, and increased connectivity and accessibility to reduce geographical isolation of the northeast peninsula of Malaysia (Lall 1996).

Incentives in the form of regulations and subsidies have also been widely used to stimulate economic growth of lagging regions. The rationale behind providing fiscal incentives is to offset costs of firm location that may arise due to various factors, such as transport and logistics costs, infrastructure conditions, factor price differentials and lower levels of public services and amenities. Tax holidays and interest rate subsidies have been extensively tried out to develop an autonomous industrial base in the Northeast of Brazil; reductions in import duties, income, sales, and capital gains taxes, and lower interest rate were used to move industry away from the largest agglomerations in Mexico; and extensive tax holidays were used as an incentive for firms in Thailand to move out of Bangkok to regional cities.

In this section, we provide a selective review of programmes undertaken by federal governments to either stimulate the development of lagging or underdeveloped regions or to deconcentrate growth out of the largest agglomerations. This review is by no means exhaustive, but is meant to illustrate the range of policy responses to address regional or spatial inequalities. Whenever possible, we have described the industrial outcomes following these interventions. It is, however, very difficult to assess the true impact of these interventions as most evaluations have not used control groups and often do not take into account methodological issues that make it difficult to identify the impact of these programmes. We first review the empirical evidence on the effects of infrastructure investments, and then discuss the performance of regional incentives.

INVESTMENTS IN INFRASTRUCTURE

There are three basic ways in which infrastructure affects productive activities and social welfare. First, infrastructure has a direct effect on regional production and employment, as it is an intermediate public good used in the production process, and infrastructure development itself increases employment. Second, infrastructure influences business and industrial location decisions as it reduces costs of production and distribution. Third, it improves welfare and 'quality of life' in the region by enhancing its amenity value as certain services derived from the use of infrastructure are used directly as final goods.

In order to assess the importance of infrastructure in promoting the growth of lagging regions, we need to address the following issues. What is the impact of infrastructure investments on productivity of existing firms? To what extent do infrastructure improvements attract new economic activity and private capital? Conventional wisdom and a host of empirical studies clearly suggest that infrastructure is a necessary condition for economic growth, and it enhances economic performance of existing industry in lagging regions. However, the evidence of the effects of infrastructure development on stimulating new economic activity in lagging areas or the extent to which these investments bridge development outcomes with leading areas is mixed.

While, in principle, infrastructure investments should help make lagging regions more attractive to investors, often these investments by themselves are not enough to compensate a firm for profit differentials from locating in an already established region. Other factors at play include the regional policy environment, accessibility to domestic and foreign markets, and the presence of other firms in the same or related industries. In this context, infrastructure improvements while being necessary, are not sufficient for the development of lagging regions.

There is abundant empirical evidence of the contribution of infrastructure in raising productivity. A revival of academic interest on examining the contribution of infrastructure followed Aschauer's (1989) work on the United States and Biehl's (1986) paper on the European Community, which used an aggregate production function approach to show that infrastructure investments had significant productivity and growth effects. In the United States, this sparked debates on whether the national productivity slowdown was attributable to reduction in infrastructure investments. Other empirical studies along these lines include Morrison and Schwartz (1996) and Nadiri and Mamuneas (1994) who find that infrastructure provision translates into cost savings. For Germany, panel estimates for 11 federal states of (West) Germany for the period 1970–88 show that public capital formation encourages private investment (Seitz and Licht 1992) and contributes positively to cost savings (Conrad and Seitz 1992).

Data from Spanish regions for the period 1964–91 also confirm that public capital (roads, water infrastructures, ports and urban structures) have a significant positive effect on value added (the estimated elasticity is 8 per cent). Interestingly, the results also show that the effects of infrastructure on growth have reduced over time (Mas *et al.* 1995), signifying that there are diminishing returns to these investments. Evidence from developing countries using the aggregate production function approach is limited, particularly due to problems in obtaining reliable data on public capital stocks. In a study of infrastructure and lagging region development in India, Lall (1999) shows that the benefits of infrastructure improvements are disproportionately felt in lagging states. These results are consistent with the motivation for infrastructure investments that increasing infrastructure stock in lagging regions, like increasing any other stock of capital will improve regional productivity.

Findings on the aggregate benefits of infrastructure are also supported by micro or firm level evidence. In principle, improved transport linkages will enhance access to consumer markets and intermediate buyers and suppliers, which will in turn increase the demand for a firm's products and provide incentives to increase scale and invest in cost-reducing technologies. In addition, availability of high quality infrastructure linking firms to urban market centres increases the probability of technology diffusion through interaction and knowledge spillovers between firms, as well as between firms and research centres, and also increases the potential for input diversity (Lall et al. 2004). Thus, improved accessibility has the effect of reducing geographic barriers to interaction, which increases specialized labour supply and facilitates information exchange, technology diffusion and other beneficial spillovers that have a self-reinforcing effect.

There is an emerging body of empirical studies which examines the links between market transport infrastructure and firm-level productivity or costs.[2] There are several studies for India that show that market accessibility is associated with higher TFP and labour productivity (Lall et al. 2004, Lall and Mengistae 2005). Similarly for Brazil, changes in transport costs to the nation's largest market (São Paulo) have statistically and economically significant impacts on firm-level costs. For example, a 10 per cent reduction in travel time to São Paulo would reduce firm-level costs by 1.6 per cent in the garments and textiles industry and by 2.3 per cent in the leather products industry (Lall et al. 2004b).

While the evidence presented here suggests that infrastructure investments enhance productivity, an equally important task is to find out if these investments stimulate new investment or job creation in lagging regions. The empirical evidence on the impact of infrastructure improvements in stimulating industrial activity to locate in lagging regions, or improve aggregate development outcomes of lagging regions is weak at best. For example, Aberle and Towara (1987) illustrate this point using data for Italian regions. Improvement in transport infrastructure via the construction of motorways was an important component of the Italian government's regional policy in its effort to reduce inequalities between the highly industrialized North and the mainly agricultural South. Principal component analysis of 15 provinces of Campania, Puglia, Basilicata and Calabria for the period 1961–80, however, fails to attribute regional economic development (as measured by net per capita GDP) to the improvements of the transport infrastructure. The limited role of transport infrastructure in this case can be attributed to the fact that regions in southern Italy (with the exception of Sardinia) have underused roads, and that development in the South was conditioned primarily by the lack of private investors and a qualified workforce. Infrastructure, in this case, was not a binding constraint.

Similarly, de la Fuente and Vives (1995) show that investments in infrastructure have only made a small contribution to regional convergence in Spain. They find that while differences in infrastructure endowments account for about one-sixth of observed regional per capita income inequality of 17 regions in Spain for the period 1980–91, the evolution of infrastructure stocks, in practice, had only a minor impact on income disparities.

Turning from these aggregate studies on convergence, we look at location decisions of individual firms and examine the extent to which location decisions are contingent on transport improvements. In principle, a firm will choose to locate production in an area where its profits will be the highest. Thus, if infrastructure provision serves to bridge inherent disadvantages of lagging regions in terms of market access, a relevant question in terms of lagging region development is the type and scale of improvements in regional characteristics that are needed to offset benefits from locating in existing agglomerations, and induce firms to move towards peripheral locations.

Deichmann et al. (2005) address this question for Indonesia, and estimate a firm-level location choice model to illustrate the potential impacts of transport improvements on relocation of firms, particularly into the lagging eastern part of the country. Their findings can be organized into two parts: (1) the effect of transport infrastructure on regional attractiveness, and (2) the implications of expanding transport capacity on the industrial prospects of eastern Indonesia.

On the first issue, that is the impact on regional attractiveness, access to export markets measured by proximity to the nearest port increases the attractiveness of a region. They find that port access is highly significant in the apparel, wood, and paper products sectors, all of which have a strong export orientation. For example, in wearing apparel, the estimated coefficient of –0.07 means that a 10 per cent reduction in distance to the nearest international port will be associated with an approximately 0.7 per cent increase in regional attractiveness or potential profitability. Similarly, internal market access also enhances regional attractiveness. The estimated coefficients for road density are positive and statistically significant for seven industry sectors, with large elasticities in the textiles, wearing apparel and furniture sectors.

To examine the implications of expanding transport capacity, they ask the following questions: What would happen to distribution of manufacturing if road densities in peripheral Eastern Indonesia were upgraded to a level similar to the country's major agglomeration (Jakarta)? Would this stimulate large-scale migration of firms to peripheral areas? Or are agglomeration forces and other amenities strong enough to limit the scale of firm migration? They find that some relocation of firms in transport sensitive sectors is likely under these circumstances, but the relocation is limited from major agglomerations to other large cities (such as Surabaya) with comparable amenities, and not to the peripheral parts of the country.

This tells us that large scale transportation improvements, without other forms of public service provision and amenity creation, may not be adequate to induce firms to relocate from agglomerations to peripheral areas. These findings are consistent with predictions in Krugman's (1991b) analytic work, where in the presence of large regional differences in local amenities (local infrastructure), improvements in inter-regional infrastructure without any concomitant improvement in lagging region 'local infrastructure' will not induce firms to relocate out of major agglomerations.

While infrastructure links may enhance productivity and attract firms to lagging regions, these investments can also have unintended consequences. Inter-regional transport investments that improve the connectivity of a lagging region implicitly reduce a natural tariff barrier. They not only provide local firms with better access to the inputs and markets of more developed regions, they also allow firms serving larger markets (such as non-basic goods producers in agglomerations) and benefiting from economies of scale and lower unit costs of production to expand into lagging region markets in competition with local producers. That is, the opposite of the intended effect may take place. In the case where manufactured goods are standardized and product substitution is relatively costless, instead of manufacturing activity moving to or being created in the lagging region, we may, in fact, see reduced industrialization where producers in the leading regions expand production and crowd out local producers. Puga (2002) puts this in the context of new economic geography models, where in the presence of limited inter-regional migration and small differences in wages even with significant variations in other regional attributes, infrastructure investments do little to bring about convergence, and may even widen disparities.

REGIONAL INCENTIVES

Regional incentives have been widely used to attract industries and stimulate the growth potential of lagging regions. The main objective of federal incentives for regional development is to reduce inter-regional differentials which may serve as an inducement for firms to locate in lagging or underdeveloped regions. Table 6.1 provides some examples of typical incentive programmes used in various countries. These include interest rate subsidies, tax holidays, industrial estate development, and business licensing regulations. The evidence on the impact of regional incentives is mixed. While incentives have been used in many developed as well as developing countries, there is no conclusive evidence to suggest that these policies have succeeded in transforming the fortunes of lagging sub-national regions.

From the instruments listed in Table 6.1, we now look more closely at the following programmes to get an insight into the design of these programmes:

(1) investment subsidies in Brazil
(2) tax and import duty exemptions in Mexico
(3) tax holidays in Thailand
(4) revenue sharing in Korea

INVESTMENT SUBSIDIES IN BRAZIL

Financial outlays from the central government in Brazil to support spatially explicit programmes have been estimated to be US $3 to 4

Table 6.1: Examples of regional incentives

Instrument	Examples		
Investment subsidies	Brazil—interest rate subsidies—FNE, FNO —constitutional funds	India—concessional finance (Fifth Five Year Plan)	Thailand—subsidized credit for locating in secondary cities
Tax holidays	Brazil: Corporate tax exemption granted to enterprises during the first 10–15 years of operation	Korea—corporate income taxes exemption for 3 years to locate in regional industrial estates	Thailand—income tax exemptions; sales tax reductions for firm locating in industrial processing zones (IPZ)
Reductions in import duties	Mexico—import duty exemption for locating outside of the three largest metro areas	Thailand— reduction in import duties on raw materials	
Industrial estates	Brazil: provision of industrial land and infrastructure	Japan: industrial estate development (industrial relocation policy 1975-80)	Thailand— Development of industrial estates
Regulation	India—preference to backward areas in industry licensing (Fifth Five Year Plan)	Korea—controls on new industrial development in Seoul	
Other programmes	Japan—wage subsidies for companies creating new employment opportunities by constructing or expanding their manufacturing facilities	Korea—sharing of tax revenues to support programmes in underdeveloped regions	Mexico—technical assistance in the form of pre-investment studies, market research, and assistance in obtaining credit.

Source: Compiled from various sources, discussed in text.

billion per annum in recent years (Ferreira 2004). The estimated cost of tax breaks and associated regional development programmes (excluding the Zona Franca de Manaus) in 2002 was estimated at almost US $900 million (Secretaria da Receita Federal 2003). Tax credits directed to the Zona Franca de Manaus are estimated to be US $1.2 billion in 2003 alone. Investment incentive programmes for the North and the Northeast, funded by income tax deductions, averaged more than 600 million dollars a year between 1995 and 2000, before they were shut down after accusations of mismanagement.

Explicit spatial policies in Brazil include three sets of instruments which target private sector development through various kinds of subsidies: (1) fiscal incentive programmes such as those administered by SUDENE (Superintendência do Desenvolvimento do Nordes), SUDAM (Superintendência de Desenvolvimento da Amazônia), and the Zona Franca de Manaus (Manaus Free Trade Zone); (2) subsidized credit channeled through the CFs (Constituicao Federal), which has become one of the most important instruments of spatial policy in Brazil; (3) and regional development banks, such as Banco do Nordeste do Brazil (BNB).

In 1989, the Brazilian Congress instituted three Constitutional Investment Funds (Fundos Constitucionais de Financiamento) for the Northeast (FNE), the Center-West (FCO), and the North (FNO). The main aim of these funds was to stimulate economic and social development in these regions by extending credit to local entrepreneurs. Preferential treatment was provided to micro- and small-scale agricultural producers and small-scale manufacturing to encourage the use of local raw materials and labour. The majority, that is, 60 per cent of the outlays for the CF's were allocated to the Northeast, and 20 per cent each were allocated to the North and the Center West. Funds are transferred from the National Treasury to the Ministry of Integration ('Ministério da Integração'), which later re-allocates them to the operating banks—the Banco do Nordeste (FNE), Banco da Amazônia (FNO) and Banco do Brasil (FCO). The CFs are financed by receipts from income taxes and taxes on industrial products.

Interest rate subsidies are the main incentive offered through the CFs. While market interest rates offered to private firms are currently more than 45 per cent, the CFs offer credit at 8.75 per cent to non-agricultural micro firms; 10 per cent to small firms, 12 per cent to medium sized firms; and 14 per cent to large enterprises. Interest rates are even lower for comparable agro producers: 6 per cent for mini-producers, 8.75 per cent for small to average and 10.75 per cent for large ones. These interest rates were negative in real terms in 2002, when inflation was 12.5 per cent. Rates offered to individual producers varied by sector, investment size, and credit record of the borrower. Between 1989 and 2002, more than US $10 billion were provided in subsidized loans, which is about 0.8 per cent of the total GDP of the 3 regions per year (Ferreira 2004).

The conclusions of these actions for policy makers are mixed. To the extent that regional interventions for reducing spatial disparities are explicit goals in themselves, we find that the CFs were successful in inducing firms to locate in Brazil's lagging regions and that the CFs may yield more 'bang for the buck' if they are used to induce entry by firms' headquarters into those regions.

TAX AND IMPORT DUTY EXEMPTIONS IN MEXICO

The Mexican Government has historically used fiscal incentives to promote industrial development outside the three largest urban agglomerations: the Mexico City Metropolitan Area (MCMA), Guadalajara, and Monterrey. Between 1970 and 1980, an elaborate tax and duty exemption system was set up. Fiscal incentives were provided to industries to locate outside the three largest urban centres. In order to provide fiscal incentives, the country was divided into three zones. Zone 1 included the largest three metropolitan areas; Zone 2, the secondary cities of Pueblo, Cuernavaca, Queretaro, and Toluca; and Zone 3, the remainder of the country. Industries locating in zones 2 and 3 were eligible for 50 to 100 per cent reduction in import duties, income, sales, and capital gains taxes, as well as accelerated depreciation and lower interest rates. In addition, firms with investments less than 5 million pesos could receive technical assistance in the forms of pre-investment studies, market research, and assistance in obtaining credit.

We have seen a pattern of re-concentration of manufacturing away from Mexico City to northern cities such as Ciudad Juarez, Monterrey, and Tijuana, which are physically close to the United States, following the opening of the Mexican economy to foreign trade and investment with NAFTA (North America Free Trade Agreement) (Hanson 1998). Since 1980, industrial activity in Mexico has moved to states on the US–Mexico border, reducing the importance of Mexico City as the nation's main industrial centre. Between 1980 and 1993, the border states increased their share of manufacturing employment from 21 per cent to 29.8 per cent, and Mexico City's share of manufacturing employment declined from 44.4 per cent to 28.7 per cent. However, reviews of the programme show that the overall impact of fiscal incentives for decentralization has been either insignificant or undesirable. The evaluations identify that taxes such as import duties on raw materials and capital goods were very low to begin with, and a further reduction had no effect on private location decisions. Thus, tax exemption may have resulted in unnecessary loss of public revenues.

TAX HOLIDAYS IN THAILAND

The Thai Board of Investment tried to increase the growth rate of regions outside Bangkok in the 1970s and 1980s by offering tax holidays to new firms. These incentives included: (1) exemptions or 50 per cent reductions of import duties and business taxes on imported machinery; (2) Reductions of import duties and business taxes of up to 90 per cent on imported raw materials; (3) corporate income tax exemption for 3–8 years; (4) exemption of up to 5 years on withholding tax on goodwill, royalties or fees remitted abroad; and (5) exclusion from income tax of dividends derived from promoted enterprises during the tax holiday. Industries establishing in the four special industrial processing zones could claim: (1) 50 per cent reduction in sales tax and corporate income tax for 5 years; (2) allowance to deduct double the cost of transportation, electricity and water supply from taxable income for eight years, and (3) deduction of up to 10 per cent of facilities up to 10 years. (More generous incentives were offered for Khon Kuen and Songkhla, with sales tax reduction raised to 75 per cent, deduction of utility costs allowed for 10 years and deductions for facilities raised to 20 per cent.)

These fiscal incentives, however, did not result in a large shift of investment from Bangkok to regional cities. The failure to achieve the dispersal of economic activity can partly be attributed to problems with the design of the incentive programme, which was in the form of deductions from taxable profits (World Bank 1980, 1986b). Producers in regional cities faced persistent cost disadvantages, and the break-even date when new firms start making profit would probably be later for these businesses. Thus, initial tax holidays in this case were not much of an inducement. Further, eligibility criteria based on a minimum size cut-off (defined in terms of minimum capital investment or capital assets or production capacity) made most small scale and agro based local firms ineligible for the Board of Investment's incentives.

REVENUE SHARING IN KOREA

Industrial development in Korea has historically been around the large agglomerations of Seoul and Pusan. In an attempt to promote balanced regional growth and divert growth away from Seoul, the Korean government initiated large scale programmes in the 1960s and 1970s. As part of this strategy, considerable resources were redistributed from major cities to less developed provinces in the form of block grants and other transfers. In terms of the central government tax burden, Seoul, Pusan, Taegu, and Inchon generate the bulk of internal taxes in Korea, with Seoul city accounting for about half the national internal taxes in 1983. Under the Local Share Tax Law (enacted in 1963), 13.2 per cent of domestic taxes are annually earmarked for block grants, and each city's allocation is based on the difference between the city's own revenue source and estimated needs based on standardized service provision assumptions. As a consequence of these policies, the dominance of Seoul and Pusan declined in the 1970s and 1980s, but growth continued to reinforce the already developed Seoul–Pusan axis—most growth occurred either just outside the boundaries of Seoul and Pusan, or in a range of small and medium-sized cities strung along the developed axis (World Bank 1986a; Murray 1988).

WHAT LIMITS THE PAY-OFF FROM SPATIALLY TARGETED INTERVENTIONS?

The evidence reviewed in the previous section suggests that regional incentives and infrastructure investments have limited pay-offs in terms of stimulating industrial growth in lagging regions. In this section, we look at what kinds of spatial policies have been enacted in India and ask what factors limit the returns from these interventions?

From Chapter 2, let us revisit some of the major spatial policy initiatives:

(1) Heavy industry was discouraged (and eventually forbidden) from locating in metropolitan centres. One of the reasons for this was the deterioration in living conditions, especially for the working classes in the larger cities, and attendant problems of slums and environmental pollution. The GOI decided that

no more licenses should be issued to new industrial units within certain limits of large metropolitan cities having a population more than 1 million and urban areas with a population of more than 500,000 in the 1971 census.[3]
(2) Large public sector projects (steel plants, for example) were located in lagging states like Madhya Pradesh (Bhilai) and Orissa (Raurkella) in the 1950s and 1960s. These effectively became growth centres (a term which did not officially exist at the time when these came into being). The growth centre policy was officially used later. For instance, in Maharashtra, Nashik became a growth centre in 1967, Roha in 1968, Nagpur in 1969, and Akola, Chadrapur, Dhule, Nanded, and Ratnagiri in 1988.
(3) Financial incentives (typically in the form of tax holidays) were provided for private industrial investment in designated lagging districts (about 60 per cent of all the districts in India). State governments also enacted their own policies to attract capital to lagging districts. For instance, West Bengal and Maharashtra created three- and five-category systems for identifying districts which districts should receive tax and duty abatements (and to what degree).

We have shown how ineffective these policies have been in recent years. Earlier, up to 1980, almost 55 per cent of the (central) capital subsidies went to only 25 out of 296 eligible lagging districts, where all 25 were in industrially advanced states. We argue that there are four primary reasons for the failure of these interventions.

(1) The natural tendency of economic activity to cluster is one of the main reasons for the limited pay-off from regionally targeted interventions that seek to stimulate industrial development in disadvantaged regions.
(2) The multilevel policy framework (at the central and state levels) is inherently contradictory at all levels—what one hand gives, the other takes away—as a result of which, it is unclear whether the targeted regions actually receive substantial advantages from spatial policies. In addition, explicit spatial policies are usually subsumed by implicit spatial policies that

apparently have no spatial dimension but often have far deeper spatial impact than the explicit ones.
(3) To improve the attractiveness of a region for specific industries, a package of interventions that address a firm's location criteria needs to be developed—uncoordinated interventions that only partly compensate for a region's 'unattractiveness' are unlikely to be successful.
(4) Regional and local institutions can become major barriers to capital flows. These institutions are historical legacies, often dating back from colonial times. At the regional level, these institutions can lead to identity-based and redistribution-oriented policies that seriously hamper private capital formation.

CLUSTERING OF ECONOMIC ACTIVITY

As described in the beginning of this chapter, the spatial distribution of economic activity is very unequal, with most modern sector industrial activity locating in a few areas within most countries. So why does industry cluster? We have argued that there are two main groups of factors external to the firm that influence industry clustering: (1) natural advantage offered by specific regions and (2) production related externalities. Some regions have a natural advantage that makes them relatively more attractive to different types of firms. This may include induced advantages such as good transport infrastructure as a result of past public investment. These factors are central to the NEG models, where firms tend to locate in areas that have high demand for the good they produce facilitated through good transport facilities and thus market access. Second, there are reasons for locating in a certain region that are more specific to the firm's production process and its interaction with suppliers, customers or competitors. These are production externalities where firms locate in proximity to other firms to accrue externalities such as knowledge/information transfers, maximize trading linkages and benefit from the presence of ancillary services which co-locate near industrial centres.

For example, in Indonesia, the dominance of Java and particularly West Java's Jabotabek region in terms of manufacturing activity derives from its natural advantage.[4] Historically, Java's fertile volcanic soils supported high population densities, and by the sixteenth century the port of Sunda Kelapa in today's Jakarta had established itself as an important trade hub. This, in turn, attracted the establishment of European trading posts, and eventually the establishment of the capital of the Republic of Indonesia. In the post-colonial period, Indonesia developed, what some have called an economic system of 'bureaucratic capitalism' (and others have called 'crony capitalism'), where a high premium was on close access to members of the government. These factors accelerated the agglomeration of economic activity near the centre of power in a highly centralized political system and resulted in the rapid growth of the manufacturing sector in the Jakarta region in the 1980s and early 1990s.

The initial benefits from natural advantage can trigger a self-reinforcing process that leads to the emergence of urban–industrial agglomerations to a point where the initial advantage responsible for the growth of the centre may no longer be important. Building on natural advantage, production externalities further enhance clustering processes. Production externalities relate to dynamics that directly affect the firm's microeconomic decision-making. Most fundamentally, clusters of firms that are predominantly in the same sector take advantage of localization and urbanization economies. These have been extensively discussed in Chapters 1, 4, and 5, and do not need to be repeated here. What needs to be reinforced is our consistent finding that localization economies play a much smaller role than urbanization economies, which in turn gives more prominence to the size and diversity of urban agglomerations as attractors of new industrial investment.

CONTRADICTORY POLICIES

Efforts at reducing regional disparities in income have played a major role in the rhetoric of India's planning practice and policy objectives. The policy of the government has been to try to ensure balanced development across regions and to gradually reduce regional disparities at the same time that growth rates at the national level are maintained at high levels. There is some contradiction between these two policies—not merely because equity-oriented goals are inherently inefficient (more on this soon), but also because many policies are poorly thought out and because national-level policies may be counteracted by regional or local-level policies.

Consider an example of a poorly thought out policy. The Freight Equalization Policy of 1956 equalized the prices for 'essential' items such as coal, steel, and cement nationwide. This effectively negated the location based advantages of regions that were rich in these resources and placed them at a disadvantage relative to regions that produced 'non-essential' items whose prices were not equalized. The affected areas were southern Bihar, western Orissa and eastern Madhya Pradesh, all among the poorest, least industrialized parts of the country; also affected was West Bengal, which, not coincidentally, went into a deep decline at about the same time that this policy went into effect. The Freight Equalization policy has been discontinued since the early 1990s, but the damage may already have been done.

Or consider the impact of national macroeconomic policies that outwardly have no spatial impact, for example, an exchange rate policy that makes imports cheaper and exports dearer. GOI followed exactly such a policy well into the 1980s. The purpose was to make the import of industrial machinery cheaper in order to enable and speed-up the industrialization process. At the same time, this policy weakened the agricultural commodity export potential of the predominantly agricultural nation. Urban regions were subsidized at the cost of rural agricultural regions, urban production was subsidized at the cost of rural production and agro-based industries (Lipton 1977). It is difficult to imagine any spatially targeted tax abatement policy that can counteract the effect of such a sweeping macro-economic policy.

Finally, consider the actions taken at the state level to boost growth and attract capital. We have noted that in addition to national level location incentives, there are often state level location incentives. However, the impact of these policies tends to get dwarfed by the impact of urban and metropolitan spatial targeting at the state level. State governments create an array of packages—special economic zones, export processing zones, information technology hubs, etc.—that receive more attention and funding, almost without exception within advanced industrial regions. Therefore, a tax policy that appears to provide incentives in lagging regions gets swamped by an infrastructure policy that increases the advantages of the leading regions.

Uncoordinated Policies

In developing the Five Year Plans, one of the Government of India's objectives has been to improve the standards of living of the lowest 30 per cent of the population, most of them living in 'backward areas' (Das 1993). Instruments such as the minimum needs programmes have been used to operationalize these objectives. The Planning Commission reported that decisions on resource transfers, agriculture development strategies, and industrial location policies were not, however, always based on a systematic analysis of regional disparities and their underlying causes. In a review of planning and regional differences, Das (1993) indicates that the Fourth Five Year Plan made the first explicit attempt since the early 1960s to review the patterns of regional development in the context of policy instruments.

While the evidence (although limited) shows that inter-regional transport improvements allow firms to relocate from core metro areas to their peripheries, there is no convincing evidence to show relocation to small urban centres in lagging areas. Experience with infrastructure investments and regional incentives suggests that these interventions, by themselves, are not sufficient to stimulate industrialization in lagging regions. In this context, it becomes important to examine what regional attributes are valued by firms, and then develop a package of interventions that improves the attractiveness of these regions.

There are two broad approaches to identifying the factors that influence industry or firm location decisions. One is to ask firms directly, as is typically done in investment climate surveys. Such surveys gather responses from firm representatives on aspects of the business environment, including the quality of utilities, infrastructure and the regulatory environment that affects a firm's operations. For example, there are often large geographic variations in the frequency of power cuts or in the time it takes to acquire a telephone line. Road quality and access to transport hubs can be measured in alternative ways, for example, by measures of road density or distance from a container port. Measures of the regulatory environment include the time taken to start a new business, the procedures for hiring and laying off employees, and the existence of corruption among officials. Such measures are of direct importance in enabling areas to benchmark their performance, as well as being inputs to a broader analysis.

A second source of evidence on the determinants of firm location decisions is econometric analysis of empirical patterns. A typical study of plant location might estimate the probability of a firm (in a specific sector) establishing production in a particular location, based on a number of explanatory variables that describe factor prices, as well as barriers and opportunities across potential locations. The variables may include: factor prices; quality and cost of complementary utility services, including electricity, water and telecommunication; market access as a function of the size of the region that can be reached given existing transport infrastructure; agglomeration economies as measured by the presence of firms in own industry and of firms in related—for example, buying or supplying—industries; labour and other regulations.

Recent evidence reported in the Investment Climate analysis for Indian regions (cities and states) shows that in addition to the presence of firms in the same or related industries, firms place considerable value on interstate variations in labour regulations, the administration of day to day business regulations, availability and quality of infrastructure, and state-level land and real estate taxes (World Bank 2004b). States where local administrators are predatory in terms of administering local business regulations are less attractive as investment destinations. Similarly, breakdowns in power supply have significant adverse effect on regional attractiveness. Complementary work on determinants of urban growth in Brazil shows that in addition to inter-regional infrastructure connectivity, cities which do better in reducing crime and violence, as well as providing local amenities and public services, have higher income and productivity growth (da Mata et al. 2005).

HISTORY AND INSTITUTIONS

Finally, we consider the reality of the persistence of social institutions and the long-term impact of institutional arrangements that are designed for expropriation. In this line of argumentation, the line of causality runs through redistribution and its growth damaging potential. More specifically, a history of expropriation is likely to give rise, eventually, to institutions that are redistribution-focused. Arthur Okun (1975) argued that redistribution is like carrying water from the rich to the poor in a 'leaky bucket'. This position has come under sustained intellectual and empirical criticism by several new growth theorists (Persson and Tabellini 1994; Alesina and Rodrik 1994; Clarke 1995). The reasons for questioning the value of redistribution as a general principle are spelled out in Fields (2001) and Chakravorty (2005) and are quite compelling. Nonetheless, it is possible, that in some (but not all, or even in a majority of) situations, the politics of redistribution becomes all-consuming. Growth becomes a decidedly secondary objective; policies oriented toward growth are viewed with suspicion and often opposed. Private capital is likely to deem such places inhospitable, fuelling a vicious (or cumulative) cycle of decline.

There is some interesting work beginning to emerge on the long-term impact of institutions of expropriation. Engerman and Sokoloff (1997) argue that institutions concerned with the diffusion of political power, namely, universal suffrage, have also a significant impact on economic growth. Population heterogeneity, in their view, is a slowing factor in the evolution of voting rights. In India, Banerjee and Iyer (2005) focus on colonial land revenue institutions and prove that in post-independence India, agricultural investment and productivity were significantly higher in districts where historically land rights were given to cultivators than in districts where rights were given to landlords. The latter were the regions of north and east India where the zamindari system was used for revenue collection, specifically in Bihar and Bengal. Their analysis of India's land tenure under the British rule and regional economic performance today illustrates the phenomenon of an institutional overhang; colonial institutions which have not existed for 60 years continue to affect the modern economy.

This brings us back to the major insight reported in Chapter 2: that in making industry location decisions, the decision makers choose the general region of location first and the specific point of location second. It is the first decision that is critical for regional development. Therefore, this is the decision that needs to be influenced. It is possible that decision makers in private industry do not carry around the whole map of India in their heads; large portions of it are masked out. Some regions are masked out because they do not have sufficient infrastructure or have poor market access. Other regions are masked out because the political wars initiated during colonialism are still being waged through parochial distributive politics.

OPTIONS FOR BIHAR

Finally, we return to our original question: what can Bihar do? In previous sections, we reviewed selected experience with the use of regional policy instruments to stimulate economic growth in spatially and historically disadvantaged regions. Our primary lesson is that public expenditures in infrastructure and targeted regional incentives, by themselves, are not adequate to transform the economic fortunes of lagging reasons. Given that we now know more about what not to do, what can we say about things that can be done to improve development outcomes in these regions?

Let us consider Bihar as an example to develop some concrete points of intervention. Bihar is arguably one of India's most backward states in terms of development indicators and has the highest poverty incidence in the country. Over 75 per cent of its workforce is occupied in agriculture. The representation of industry has become particularly low following the bifurcation of the state with most of its industry and mining enterprises now being in the new state of Jharkhand. Agriculture represents 39 per cent of the state's GDP and the industry sector contributes only 12 per cent (the lowest among India's major states), employing only 10 per cent of the workforce (World Bank 2004a). The performance of agriculture, however, has been dismal with a negative growth in agricultural output per capita between the mid-1990s and 2001. In addition, excessive volatility in agricultural output has had adverse implications on livelihoods for the majority of the state's population.

In this context, what should regional development policies for Bihar look like? As agriculture is likely to be a prominent feature of Bihar's economy for the foreseeable future, reducing the volatility of agricultural output and increasing its productivity should be key priorities. Investments in infrastructure that improve local conditions, such as electric power and irrigation, along with measures for flood management are likely to produce considerable benefits as they will strengthen the performance of the mainstay of the local economy.

Turning to industrial development, it does not appear that standardized manufacturing development is likely to be successful, especially if production of these standardized products is already concentrated or clustered in other parts of the country. However, development of an agro-based industrial strategy may be worth considering, as it would stimulate the development of upstream and downstream industries. This, coupled with specific investments in inter-regional infrastructure may create new local and regional markets. While some degree of fiscal support through a variety of incentives may be required at various points, these should be outcome or performance related grants. Incentives should be linked to their contribution to new employment, and improvements in local production and the State Domestic Product.

In conclusion, we would like to play devil's advocate in raising a possibility that seems outside the pale. From the beginning of this chapter, we have looked at only one development alternative—bringing jobs to people, or development in place. The alternative is, of course, bringing people to jobs. This migration based strategy has a long history, beginning with the 'out of Africa' movements into the rest of the world, through the settlement of north India by migrants from the Caucasus. In the industrial era, this has led to the settlement patterns we see today; large coastal metropolitan regions, overrun, in the developing world, with slums, and accused of parasitism and unsustainability. Every developing state would like to curb the growth of these giant settlements; every socialist state has actively tried to do so.

We ask the question: is it possible to envision better planned metropolitan regions that are not only larger than the ones we see today, but are ethnically and industrially diverse, and do a better job of enhancing overall welfare than the existing system does? If migration is a natural response to development differentials, should we not recognize and deal with it? When development in place is not possible, why not encourage migration? We write this fully recognizing the value of place, of culture, memory, and tradition, and the out-group hostility faced by ethnic 'others'. We merely suggest that gigantic metropolitan size is not the end of the world. There are technological solutions for urban size. Metropolitan Tokyo's 25 million people use these technological solutions everyday. Some creative thinking on metropolitan planning and management may go a longer way toward societal welfare enhancement than have the traditional growth management and inter-regional development approaches.

NOTES

1. We can also think of investments in social infrastructure, but these have typically not been used as explicit instruments of regional policy. Thus, we limit our discussion to fiscal incentives and public expenditures on physical infrastructure.
2. These studies are limited in focus to manufacturing industry
3. Details on this policy can be obtained at: http://www.smallindustryindia.com/policies/iip.htm#Indus3
4. Jabotabek is the metropolitan area consisting of Jakarta, Bogor, Tangerang, and Bekasi.

References

Abdel-Rahman, H. 1988. 'Product Differentiation, Monopolistic Competition, and City Size', *Regional Science and Urban Economics*, 18, pp. 69–86.

Aberle G. and M. Towara. 1987. 'Transport Infrastructure and Regional Development in the South of Italy: A Macroeconomic Approach', *Journal of Regional Policy*, 7, pp. 421–30.

Agresti, A. 1996. *An Introduction to Categorical Data Analysis*. New York: John Wiley.

Ahluwalia, I. J. 1991. *Productivity and Growth in Indian Manufacturing*. New Delhi: Oxford University Press.

Ahuja, V., B. Bidani, F. Ferreira, and M. Walton. 1997. *Everyone's Miracle? Revisiting Poverty and Inequality in East Asia*. Washington, DC: World Bank.

Akita, T. and R. A. Lukman. 1995. 'Interregional Inequalities in Indonesia: A Sectoral Decomposition Analysis for 1975–92', *Bulletin of Indonesian Economic Studies*, 31, pp. 61–81.

Alesina, A. and D. Rodrik. 1994. 'Distributive Politics and Economic Growth', *Quarterly Journal of Economics*, 108, pp. 465–90.

Alonso, W. 1980. 'Five Bell Shapes in Development', Papers of the Regional Science Association, 45, pp. 5–16.

———. 1964. *Location and Land use*. Cambridge, Mass: Harvard University Press.

Amin A. 2002. 'Industrial Districts', in E. Sheppard and T. J. Barnes (eds). *A Companion to Economic Geography*, pp. 149–68, London: Blackwell.

Amirahmadi, H. and W. Wu. 1994. 'Foreign Direct Investment in Developing Countries', *The Journal of Developing Areas*, 28, pp. 167–90.

Amiti, M. and B. Smarzynska Javorcik. 2005. 'Trade Costs and Location of Foreign Firms in China'. Mimeo, Washington DC: World Bank.

Amiti, M., and L. A. Cameron. 2004. 'Economic Geography and Wages', Centre for Economic Policy Research (CEPR) Discussion Paper, London, UK, 4234.

REFERENCES

Andrade, E., M. Laurini, R. Medalozzo, and P. L. Pereira. 2004. 'Convergence Clubs among Brazilian Municipios', *Economic Geography*, 83, pp. 179–84.
Anselin, L. 1995. 'Local Indicators of Spatial Association: LISA', *Geographical Analysis*, 27, pp. 93–115.
——. 1992. *Spacestat Tutorial*. Regional Research Institute. West Virginia University: Morgantown, West Virginia.
Armstrong, H. 1995. *Trends and Disparities in Regional GDP per capita in the European Union, United States and Australia*. Brussels: European Commission.
Arrow, K. J. 1962. 'The Economic Implications of Learning by Doing', *Review of Economic Studies* 29, pp. 155–73.
Aschauer, D. A. 1989. 'Is Public Expenditure Productive?', *Journal of Monetary Economics* 23, pp. 177–200.
Aswicahyono, H. H., K. Bird, and H. Hill. 1996. 'What Happens to Industrial Structure When Countries Liberalize? Indonesia Since the Mid-1980's'. *Journal of Development Studies* 32, pp. 340–63.
Awasthi, D. N. 1991. *Regional Patterns of Industrial Growth in India*. New Delhi: Concept Publishing.
Azzoni, C. R., N. Menezes-Filho, T. A. de Menezes, and R. Silveira-Neto. 2000. *Geography and Income Convergence Among Brazilian States*. Inter-American Development Bank, New York, Research Network Working Paper R-395. Mimeo.
Baer, W. 1995. *The Brazilian Economy: Growth and Development*, 4th edition. Westport: Praeger.
Bernstein, J. I. 1988, 'Costs of Production, intra- and inter industry R & D spillovers: Canadian evidence', *Canadian Journal of Economics*, 21, pp. 324–47.
Bagchi, A. K. 1976. 'De-Industrialization in India in the Nineteenth Century: Some Theoretical Implications', *Journal of Development Studies*, 12, pp. 135–64.
Banerjee, A. and L. Iyer. 2005. 'History, Institutions and Economic Performance: The Legacy of Colonial Land Tenure Systems in India', *American Economic Review*, 95, pp. 1190–213.
Baran, P. 1957. *The Political Economy of Growth*. New York: Monthly Review Press.
Barro, R. J. and X. Sala-i-Martin. 1995. *Economic Growth*. New York: McGraw Hill.
——. 1992. 'Convergence', *Journal of Political Economy*, 100, pp. 223–51.
Bartik, T. J. 1994. 'Jobs, Productivity and Local Economic Development: What Implication does Economic Research have for the Role of the Government', *National Tax Journal*, 47, pp. 847–62.

———. 1991. *Who Benefits from State and Local Economic Development Policies?* Kalamazoo, MI: W.P. Upjohn Institute for Economic Research.

———. 1985. 'Business Location Decisions in the United States: Estimates of the Effects of Unionization, Taxes, and Other Characteristics of States', *Journal of Business and Economic Statistics*, 3, pp. 14–22.

Becker, C. M., J. G. Williamson, and E. S. Mills. 1992. *Indian Urbanization and Economic Growth since 1960*. Baltimore: The Johns Hopkins University Press.

Bernstein, J. I. and M. I. Nadiri. 1988. 'Inter industry R & D spillovers, rates of return and production in high-tech industries', *American Economic Review Papers and Proceedings*, 78: pp. 429–34.

Biehl, D. 1986. 'The Contribution of Infrastructure to Regional Development'. Mimeo, Brussels: Regional Policy Division, European Communities.

Borts, G. H. and J. L. Stein. 1964. *Economic Growth in a Free Market*. New York: Columbia University Press.

Bostic, R. 1997. 'Urban Productivity and Factor Growth in the Late 19th Century', *Journal of Urban Economics*, 4, pp. 38–55.

Boudeville, J. R. 1966. *Problems of Regional Economic Planning*. Edinburgh: Edinburgh University Press.

Bourdieu, P. 2002. 'Structures, Habitus, Practices', in C. Calhoun, J. Gerteis, J. Hoody, S. Pfaff, and I. Virk (eds), *Contemporary Sociological Theory*, pp. 276–88, Oxford: Blackwell.

Bradshaw, Y. W. 1987. 'Urbanization and Underdevelopment: A Global Study of Modernization, Urban Bias and Economic Dependency', *American Sociological Review*, 52, pp. 224–39.

Brohman, J. 1995. 'Postwar Development in the Asian NICs: Does the Neoliberal Model Fit Reality', *Economic Geography*, 72, pp. 107–30.

Brooks, D. H. 1987. 'Industrial Location and Decentralization Policies in Developing Countries'. Unpublished Ph. D. thesis. Providence, Rhode Island: Brown University.

Byres, T. J. (ed.) 1997. *The State, Development Planning and Liberalisation in India*. New Delhi: Oxford University Press.

Calzonetti, F. J. and R. T. Walker. 1991. 'Factors Affecting Industrial Location Decisions: A Survey Approach', in H. W. Herzog and A. M. Schlottmann (eds), *Industry Location and Public Policy*, Knoxville: University of Tennessee Press, pp. 221–40.

Cárdenas, M. and A. Pontón. 1995. 'Growth and Convergence in Colombia: 1950–1990'. *Journal of Development Economics* 47: 5–37.

Carvalho, A., S. V. Lall, and C. Timmins. 2005. 'Regional Subsidies and Industrial Prospects of Lagging Regions'. Mimeo, Washington DC: World Bank.

Chai, J. C. H. 1996. 'Divergent Development and Regional Income Gap in China', *Journal of Contemporary Asia*, 26, pp. 46–58.
Chakravarty, S. 1979. 'On the Question of Home Market and Prospects for Indian Growth', *Economic and Political Weekly*, Special Issue No. 4, pp. 30–2.
Chakravorty, S. 2005. *Fragments of Inequality: Social, Spatial, and Evolutionary Analyses of Income Distribution*. New York and London: Routledge.
——. 2003. 'Industrial Location in Post-Reform India: Patterns of Interregional Divergence and Intraregional Convergence', *Journal of Development Studies*, 40, pp. 120–52.
——. 2000. 'How Does Structural Reform Affect Regional Development? Resolving Contradictory Theory With Evidence from India', *Economic Geography*, 76, pp. 367–94.
——. 1996. 'Too Little, in the Wrong Places? The Mega City Programme and Efficiency and Equity in Indian Urbanization', *Economic and Political Weekly*, 31, pp. 2565–72.
——. 1994. 'Equity and the Big City', Economic Geography, 70, pp. 1–22.
Chand, M. and V. K. Puri. 1983. *Regional Planning in India*. New Delhi: Allied Publishers.
Chapman, K. and D. F. Walker. 1991. *Industrial Location: Principles and Policies*. Oxford: Basil Blackwell.
Chinitz, B. 1961. 'Contrasts in Agglomeration: New York and Pittsburgh', *American Economic Review*, 51, pp. 279–89.
Christaller, W. 1966. *Central Places in Southern Germany*, translated by C. W. Baskin. Englewood Cliffs: Prentice-Hall (originally published in 1933).
Ciccone, A. and R. Hall. 1996. 'Productivity and the Density of Economic Activity', *American Economic Review*, 86, pp. 54–70.
Clark, G. L. 1998. 'Stylized Facts and Close Dialogue: Methodology in Economic Geography', *Annals of the Association of American Geographers*, 88, pp. 73–87.
Clarke, G. R. G. 1995. 'More Evidence on Income Distribution and Growth', *Journal of Development Economics*, 47, pp. 403–27.
CMIE (Centre for Monitoring Indian Economy). 1998. *Infrastructure*. Banglore: CMIE.
CMIE (Centre for Monitoring Indian Economy). 1993. *Profiles of Districts*. Mumbai: CMIE.
Conrad, K. and H. Seitz. 1992. 'The 'Public Capital' Hypothesis: The Case of Germany', *Recherches Economiques de Louvain*, 58, pp. 309–27.
Cook, P. and D. Hulme. 1988. 'The Compatibility of Market Liberalization and Local Economic Development Strategies', *Regional Studies*, 22, pp. 221–31.

Dökmeci, V. and L. Berköz. 1994. 'Transformation of Istanbul from a Monocentric to a Polycentric City', *European Planning Studies*, 2, pp. 193–205.

D'Monte, D. 2000. *Ripping the Fabric: The Decline of Mumbai and its Mills*. New Delhi: Oxford University Press.

da Mata, D., U. Deichmann, J. V. Henderson, S. V Lall and H. Wang. 2005. 'Determinants of City Growth in Brazil'. National Bureau Economic Research (NBER) Working Paper 11585. August, Cambridge, Mass.

Daniere, A. 1996. 'Growth, Inequality and Poverty in South-east Asia: The Case of Thailand', *Third World Planning Review*, 18, pp. 373–96.

Darwent, D. F. 1975. 'Growth Poles and Growth Centers in Regional Planning: A Review', in J. Friedmann and W. Alonso (eds), *Regional Policy: Readings in Theory and Applications*. Cambridge, Mass: The MIT Press, pp. 539–65.

Das, K. 1993. 'Planning and Regional Differentiation in India: Strategies and Practices', *Journal of Indian School of Political Economy*, 5, pp. 603–32.

Das, S. K. and A. Barua. 1996. 'Regional Inequalities. Economic Growth and Liberalisation: A Study of the Indian Economy', *Journal of Development Studies*, 32, pp. 364–390.

de la Fuente, A. 2000. 'Infrastructures and Productivity: A Survey'. Mimeo, Barcelona: Instituto de Analisis Economico, CSIC.

de la Fuente, A. and X. Vives. 1995. 'Infrastructure and Education as Instruments of Regional Policy: Evidence from Spain'. *Economic Policy: A European Forum*, 10, pp. 11–51.

de Mello, L. R. Jr. 1997. 'Foreign Direct Investment in Developing Countries and Growth: A Selective Survey', *Journal of Development Studies*, 34, pp. 1–34.

Dehejia, J. 1993. 'Economic Reforms: Birth of an "Asian Tiger"', in P. Oldenburg (ed.), *India Briefing*, Boulder: Westview, pp. 75–102.

Deichmann, U., K. Kaiser, S. V. Lall, and Z. Shalizi. 2005. 'Agglomeration, Transport and Regional Development in Indonesia'. World Bank Policy Research Working Paper 3477, World Bank.

Deno, K. T. 1988. 'The Effect of Public Capital on U.S. Manufacturing Activity: 1970 to 1978', *Southern Economic Journal*, 55, pp. 400–11.

Diniz, C. C. 1994. 'Polygonized Development in Brazil: Neither Decentralization nor Continued Polarization', *International Journal of Urban and Regional Research*, 18, 293–314.

Dixit, A. K., and J. E. Stiglitz. 1977. 'Monopolistic Competition and Optimum Product Diversity', *American Economic Review*, 67, pp. 297–308.

Doms, M. E. 1992. 'Essays on Capital Equipment and Energy Technology in the Manufacturing Sector'. Unpublished Ph.D. Dissertation, Madison: University of Wisconsin.

Duranton, G. and D. Puga 1999. 'Nursery Cities'. Mimeo, London: London School of Economics.

Elizondo, R. L. and P. Krugman. 1992. *'Trade Policy and the Third World Metropolis'*. Natioanl Bureau of Economic Research (NBER) Working Paper No. 4238, NBER, Cambridge, Mass.

Ellison, G. and E. L. Glaeser. 1999. 'The Geographic Concentration of Industry: Does Natural Advantage Explain Agglomeration?', *American Economic Review, Papers and Proceedings*, 89, pp. 311–16.

———. 1997. 'Geographic Concentration in U.S. Manufacturing Industries: A Dartboard Approach', *Journal of Political Economy*, 105, pp. 889–927.

Engerman, S. L. and K. L. Sokoloff. 1997. 'Factor Endowments, Institutions, and Differential Paths of Economic Growth Among New World Economies: A View from Economic Historians of the United States', in S. Haber (ed.), *How Latin America Fall Behind*, Stanford: Stanford University Press, pp. 260–304.

Evennet, S. J. and W. Keller. 2002. 'On Theories Explaining the Success of the Gravity Equation', *Journal of Political Economy*, 110, pp. 281–316.

Expert Group on the Commercialisation of Infrastructure Projects. 1996. *The India Infrastructure Report: Policy Imperatives for Growth and Welfare*. New Delhi: National Council of Applied Economic Research.

Fan, C. C. 1995. 'Of Belts and Ladders: State Policy and Uneven Regional Development in Post-Mao China', *Annals of the Association of American Geographers*, 85, pp. 421–49.

Ferreira, A. H. B. 2000. 'Convergence in Brazil: Recent Trends and Long Run Prospects', *Applied Economics*, 32, pp. 479–89.

Ferreira, P. C. 2004. *Regional Policy in Brazil: A Review*. Report Prepared for the World Bank, Washington D.C.

Feser, E. J., and E. M. Bergman. 2000. 'National Industry Cluster Templates: A Framework for Applied Regional Cluster Analysis', *Regional Studies*, 34, pp. 1–20.

Fields, G. A. 2001. *Distribution and Development: A New Look at the Developing World*. Cambridge, Mass.: The MIT Press.

Frank, A. G. 1996. 'India in the World Economy, 1400-1750', *Economic and Political Weekly*, 27 July, pp. 51–64.

———. 1967. *Capitalism and Underdevelopment in Latin America*. New York: Monthly Review Press.

Friedmann, J. 1966. *Regional Development Policy: A Case Study of Venezuela*. Cambridge, Mass: MIT Press.

———. 1973. 'A Theory of Polarized Development', in J. Friedmann (ed.) *Urbanization, Planning and National Development*, Beverly Hills: Sage Publications, pp. 41–64.

Fujita, M. 1988. 'A Monopolistic Competition Model of Spatial Agglomeration: Differentiated Product Approach', *Regional Science and Urban Economics*, 18, pp. 87–124.

Fujita, M., P. Krugman, and A. J. Venables. 1999. *The Spatial Economy: Cities, Regions, and International Trade*. Cambridge, Mass.: The MIT Press.

Fujita, M. and P. Krugman. 1995. 'When is the Economy Monocentric? Von Thunen and Chamberlain United', *Regional Science and Urban Economics* 25, pp. 505–28.

Fujita, M. and J.-F. Thisse. 1996. 'Economics of Agglomeration', *Journal of the Japanese and International Economies*, 10, pp. 339–78.

Fujita, M. and T. Mori. 1996. 'The Role of Ports in the Making of Major Cities: Self-agglomeration and Hubb-effect', *Journal of Development Economics*, 49, pp. 93–120.

Garcia-Mila, T. and T. McGuire. 1993. 'Industrial Mix as a Factor in the Growth and Variability of State's Economies', *Regional Science and Urban Economics*, 23, pp. 731–48.

Getis, A. and J. K. Ord. 1992. 'The Analysis of Spatial Association by Use of Distance Statistics', *Geographical Analysis*, 24, pp. 189–206.

Ghosh, B., S. Marjit, and C. Neogi. 1998. 'Economic Growth and Regional Divergence in India: 1960 to 1995'. Mimeo,

Gilbert, A. 1993. 'Third World Cities: The Changing National Settlement System', *Urban Studies*, 30, pp. 721–40.

Ginsburg, N., B. Koppel and T. G. McGee (eds). 1991. *The Extended Metropolis: Settlement Transition in Asia*. Honolulu: University of Hawaii Press.

Glaeser, E. L., H. D. Kallal, J. Scheinkman, and A. Shleifer. 1992. 'Growth in Cities', *Journal of Political Economy*, 100, pp. 1126–52.

Goldar, B. 1997. 'Econometrics of Indian Industry', in K. L. Krishna (ed.), *Econometric Applications in India*, New Delhi: Oxford University Press.

Goldsmith, W. and R. Wilson. 1991. 'Poverty and Distorted Industrialization in the Northeast', *World Development*, 19, pp. 435–55.

Government of India. 1997. *Handbook of Industrial Policy and Statistics 1997*. New Delhi: Ministry of Industry.

———. 1995. *Economic Survey: 1994–95*. New Delhi: Ministry of Finance.

Greene, W. 1997. *Econometric Analysis*. Upper Saddle River, NJ: Prentice-Hall.

Greenhut, J. and M. L. Greenhut 1975. 'Spatial Price Discrimination, Competition and Locational Effects', *Economica*, 42, pp. 401–19.
Habib, I. 1975. 'Colonization of the Indian Economy: 1757–1900', *Social Scientist*, 32: 23–53.
Hansen, N. 1990. 'Impacts of Small and Intermediate-Sized Cities on Population Distribution: Issues and Responses', *Regional Development Dialogue*, 11, pp. 60–76.
Hansen, N. M. 1967. 'Development Pole Theory in a Regional Context', *Kyklos*, 20, pp. 709–27.
Hansen, W. G. 1959. 'How Accessibility Shapes Land Use', *Journal of American Institute of Planners*, 25, pp. 73–6.
Hanson, Gordon H. 1998. 'Market Potential, Increasing Returns and Geographic Concentration'. NBER Working Paper, Number 6429. Cambridge, MA: National Bureau of Economic Research.
Hanushek, E. A. and B. N. Song. 1978. 'The Dynamics of Postwar Industrial Location', *The Review of Economics and Statistics*, 60, pp. 515–22.
Harrington, J. W. and B. Warf. 1995. *Industrial Location: Principles, Practice, and Policy*. London and New York: Routledge.
Harvey D. 1982. *Limits to Capital*. Oxford: Basil Blackwell.
Head, K. and J. Ries 1996. 'Inter-City Competition for Foreign Investment: Static And Dynamic Effects of China's Incentive Areas', *Journal of Urban Economics*, 40, pp. 38–60.
Henderson, J. V. 1988. *Urban Development: Theory, Fact, and Illusion*. New York: Oxford University Press.
——. 1982. 'The Impact of Government Policies on Urban Concentration', *Journal of Urban Economics*, 9, pp. 64–71.
Henderson, J. V. and A. Kuncoro. 1996. 'Industrial Centralization in Indonesia', *World Bank Economic Review*, 10, pp. 513–40.
Henderson, J. V., A. Kuncoro, and D. Nasution. 1996. 'The Dynamics of Jabotabek Development', *Bulletin of Indonesian Economic Studies*, 32, pp. 71–95.
Henderson, J. V., T. Lee, and Y. J. Lee. 2001. 'Scale Externalities in Korea', *Journal of Urban Economics*, 49, pp. 479–504.
Henderson, J. V., A. Kuncoro, and M. Turner. 1995. 'Industrial Development in Cities', *Journal of Political Economy*, 103, pp. 1067–90.
Henderson, J. V., Z. Shalizi, and A. Venables. 2001. 'Geography and Development'. *Journal of Economic Geography*, 1, pp. 81–105.
Higgins, B. and D. J. Savoie. 1997. *Regional Development Theories and their Application*. New Brunswick and London: Transaction Publishers.
Hirschman, A. O. 1958. *The Strategy of Economic Development*, New Haven: Yale University Press.

Hodgson, G. M. 1998. 'The Approach of Institutional Economics', *Journal of Economic Literature*, 36, pp. 166–92.
Hoover, E. M. 1948. *The Location of Economic Activity*. New York: McGraw Hill.
Hotelling, H. 1929. 'Stability in Competition', *Economic Journal*, 39, pp. 41–57.
Indian Institute of Public Opinion. 1993. 'Economic Disparities Between States of the Indian Union 1951–89'. *Quarterly Economic Report*, 36, pp. 30–8.
Isard, W. 1975. *Introduction to Regional Science*. New York: Prentice-Hall.
Isard, W. 1956. *Location and Space-Economy*. Cambridge: MIT Press.
Jacobs, J. 1969. *The Economy of Cities*. New York: Vintage.
Johnson, E. A. J. 1970. *The Organization of Space in Developing Countries*. Cambridge, Mass: Harvard University Press.
Joshi, V. and I. M. D. Little. 1997. *India's Economic Reforms 1991–2001*. New Delhi: Oxford University Press.
Kaldor, N. 1970. 'The Case for Regional Policies', *Scottish Journal of Political Economy*, 17, pp. 337–48.
Kashyap, S. P. 1979. 'Regional Planning in a Consistency Framework'. Monograph Series No. 6 Ahmedabad: Sardar Patel Institute of Economic and Social Research.
Kessides, C. 1993. 'The Contributions of Infrastructure to Economic Development'. World Bank Discussion Paper No. 213, Washington, DC: World Bank.
Kieschnick, M. 1981. 'Taxes and Growth — Business Incentives and Economic Development'. *Studies in Development Policy*, Washington DC: Council of State Planning Agencies.
Kohli, A. 1989. 'Politics of Economic Liberalization in India', *World Development*, 17, pp. 305–328.
Koo, J. 2002. 'Agglomeration Revisited'. Paper presented at the Southern Regional Science Association Conference, Arlington, VA.
Kosambi, M. and J. E. Brush 1988. 'Three Colonial Port Cities in India', *Geographical Review*, 78, pp. 32–47.
Krugman, P. 1998. 'What's New about the New Economic Geography', *Oxford Review of Economic Policy*, 14, pp. 7–17.
———. 1996. 'Urban Concentration: The Role of Increasing Returns and Transport Costs', *International Regional Science Review*, 19, pp. 5–30.
———. 1993. 'The Current Case for Industrial Policy', in D. Salvatore (ed.), *Protectionism and World Welfare*, Cambridge: Cambridge University Press, pp. 160–79.
———. 1991a. *Geography and Trade*. Cambridge, Mass.: MIT Press.

REFERENCES

———. 1991b. 'Increasing Returns and Economic Geography', *Journal of Political Economy*, 99, pp. 483–99.

Kuznets, S. 1955. 'Economic Growth and Income Inequality', *American Economic Review*, 45, pp. 1–28.

Lal, D. 1995. 'India and China: Contrasts in Economic Liberalization', *World Development*, 23, pp. 1475–94.

Lall S. V. and C. Rodrigo. 2001. 'Perspectives On the Sources of Heterogeneity in Indian Industry', *World Development*, 29, pp. 2127–43.

Lall, S. V. 1999. 'The Role of Public Infrastructure Investments in Regional Development: Experience of Indian States', *Economic and Political Weekly*, 34, 12, pp. 717–25.

———. 1996. 'A Review of Economic Analysis for World Bank Infrastructure Projects'. Prepared for the Policy Research Department, World Bank: Washington DC.

Lall, S. V. and Z. Shalizi. 2003. 'Location and Growth in the Brazilian Northeast'. *Journal of Regional Science*, 43, pp. 1–19.

Lall, S. V. and T. Mengistae. 2005. 'Business Environment and Location of Industry in India'. Mimeo, World Bank.

Lall, S. V. and S. Chakravorty. 2005. 'Industrial Location and Spatial Inequality: Theory and Evidence from India', *Review of Development Economics*, 9, pp. 47–68.

Lall, S.V. and T. Mengistae. 2003. 'The Impact of Business Environment and Economic Geography on Plant Level Productivity: An Analysis of Indian Industry', Policy Research Working Paper No. 3664. World Bank, Washington DC.

Lall, S. V., S. Chakravorty, and J. Koo. 2003. 'Diversity Matters: The Economic Geography of Industry Location in India', World Bank Policy Research Working Paper 3072.

Lall, S. V., Z. Shalizi and U. Deichmann. 2004a. 'Agglomeration Economies and Productivity in Indian Industry', *Journal of Development Economics*, 73, pp. 643–73.

Lall, S. V., R. Funderburg, and T. Yepes. 2004b. 'Location, Concentration, and Performance of Economic Activity in Brazil'. Policy Research Working Paper 3268, World Bank, Washington DC.

Lee, K. S. 1989. *The Location of Jobs in a Developing Metropolis: Patterns of Growth in Bogotá and Cali, Colombia*. New York: Oxford University Press.

———. 1987. 'Employment Location and Spatial Policies: Colombia and Korea', in G. S. Tolley and V. Thomas (eds), *The Economics of Urbanization and Urban Policies in Developing Countries*, pp. 94–105, Washington, DC: The World Bank.

Leinbach, T. R. 1995. 'Regional Science and the Third World: Why Should we be Interested? What Should we do?', *International Regional Science Review*, 18, pp. 201–9.

Lewis, J. P. 1995. *India's Political Economy: Governance and Reform*. New Delhi: Oxford University Press.

Lipton, M. 1977. *Why Poor People Stay Poor: A Study of Urban Bias in World Development*. Cambridge, Mass: Harvard University Press.

Lo, F. C. and K. Salih. 1981. 'Growth Poles, Agropolitan Development and Polarization Reversal: The Debate and Search for Alternatives', in W. Stohr and J. Taylor (eds), *Development From Above or Below? The Dialectics of Regional Planning in Developing Countries*, Chichester: John Wiley & Sons, pp. 123–52.

Losch, A. 1954. *The Economics of Location*, translated by W. H. Woglom and W. F. Stolper. New Haven: Yale University Press (originally published in 1938).

Lucas, R. E. 1988. 'On the Mechanics of Economic Development', *Journal of Monetary Economics*, 22, pp. 3–42.

Markusen, A. 1995. 'Interaction Between Regional and Industrial Policies: Evidence from Four Countries', in M. Bruno and B. Plesovic (eds), *Proceedings of the World Bank Annual Conference on Development Economics, 1994*, Washington D.C.: World Bank, pp. 279–98.

Markusen, A., P. Hall, S. Campbell, and S. Deitrick. 1991. *The Rise of the Gunbelt: The Military Remapping of Industrial America*. New York: Oxford University Press.

Marshall, A. 1919. *Industry and Trade*. London: Macmillan.

———. 1890. *Principles of Economics*. London: Macmillan (8th edition published in 1920).

Martin, R. and P. Sunley. 2003. 'Deconstructing Clusters: Chaotic Concept or Policy Panacea', *Journal of Economic Geography*, 3, pp. 5–35.

———. 1998. 'Slow Convergence? The New Endogenous Growth Theory and Regional Development', *Economic Geography*, 74, pp. 201–27.

Mas, M., J. Maudos, F. Perez, and E. Uriel. 1995. 'Infrastructures and Productivity in the Spanish Regions', *Regional Studies*, 30, pp. 641–9.

Massey, D. 1984. *Spatial Divisions of Labour: Social Structures and the Geography of Production*. London: Macmillan.

McCann, P. 1998. *The Economics of Industrial Location: A Logistics-Costs Approach*. Series on Advances in Spatial Science, Berlin: Springer.

Mills, E. S. 1987. 'Non-urban Policies as Urban Policies', *Urban Studies*, 24, pp. 245–54.

Mills, E. S. and C. M. Becker. 1986. *Studies in Indian Urban Development*. Washington, DC: World Bank.

Miracky, W. 1995. 'Cities and the Product Cycle'. Unpublished Ph.D. Dissertation. Cambridge, Mass: MIT.
Mitra, A. 1965. *Levels of Regional Development in India*. New Delhi: Census of India, 1961.
ML Infomap, 1998. *IndiaMap*. New Delhi: ML Infomap Pvt. Ltd.
Mohan, R. 1983. 'India: Coming to Terms with Urbanization', *Cities*, 1, pp. 46–58.
Morrison, C. and A. E. Schwartz. 1996. 'Public Infrastructure, Private Input Demand and Economic Performance in New England Manufacturing', *Journal of Business and Economic Statistics*, 14, pp. 91–102.
Mossi, M. B., P. Aroca, I. Fernandez, and C. R. Azzoni. 2003. 'Growth Dynamics and Space in Brazil', *International Regional Science Review*, 26, pp. 393–418.
Mueller, E. and J. N. Morgan. 1962. 'Location Decisions of Manufacturers. Papers and Proceedings of the Seventy-Fourth Annual Meeting of the American Economic Association', *American Economic Review*, 52, pp. 204–17.
Murray, M. 1988. *Subsidising Industrial Location: A Conceptual Framework with Application to Korea*. Baltimore: The Johns Hopkins University Press.
Myrdal, G. 1957. *Economic Theory and Underdeveloped Regions*. London: Duckworth.
Nadiri, M. I and T. P. Mamuneas. 1994. 'The Effects of Public Infrastructure and R&D Capital on the Cost Structure and Performance of U.S. Manufacturing Industries', *Review of Economics and Statistics*, 76, pp. 22–37.
Nadvi, K. and H. Schmitz. 1999. Industrial Clusters in Developing Countries', *World Development*, 27, pp. 1503–704.
National Commission on Urbanization. 1988. *Report*. New Delhi: Government of India.
Naughton, B. 1988. 'The Third Front: Defense Industrialization in the Chinese Interior', *The China Quarterly*, 115, pp. 351–86.
Nayar, B. R. 1998. 'Business and India's Economic Policy Reforms', *Economic and Political Weekly*, 33, pp. 2453–68.
North, D. C. 1975. 'Location Theory and Regional Economic Growth', in J. Friedmann and W. Alonso (eds), *Regional Policy: Readings in Theory and Application*. Cambridge, Mass: MIT Press, pp. 332–47.
Nurske, R. 1953. *Problems of Capital Formation in Underdeveloped Countries*. London: Basil Blackwell.
Okun, A. M. 1975. *Equality and Efficiency: The Big Trade Off*. Washington, D.C.: The Brookings Institution.

Parr, J. B. 2002. 'Agglomeration Economies: Ambiguities and Confusions', *Environment and Planning A*, 34, pp. 717–31.

Perroux, F. 1950. 'Economic Space: Theory and Applications', *Quarterly Journal of Economics*, 64, pp. 89–104.

Persson, T. and G. Tabellini. 1994. 'Is Inequality Harmful for Growth?' *American Economic Review*, 84, pp. 600–21.

Petrakos, G. C. 1992. 'Urban Concentration and Agglomeration: Re-examining the Relationship', *Urban Studies*, 29, pp. 1219–30.

Piore, M. J. and C. F. Sabel. 1984. *The Second Industrial Divide: Possibilities for Prosperity*. New York: Basic Books.

Porter, M. 1996. 'Competitive Advantage, Agglomeration Economies, and Regional Policy', *International Regional Science Review*, 19, pp. 160–76.

———. 1990. *The Competitive Advantage of Nations*. London: Macmillan.

Puga, D. 2002. 'European Regional Policies in Light of Recent Location Theories', *Journal of Economic Geography*, 2, pp. 373–406.

Rajan, S. I. and P. Mohanchandran. 1998. 'Infant and Child Mortality Estimates—Part I', *Economic and Political Weekly*, Special Statistics, 9 May, 34, pp. 1120–40.

Rao, M. G. and F. Vaillancourt. 1994. 'Subnational Tax Disharmony in India: A Comparative Perspective'. National Institute of Public Finance and Policy working paper no. 4/94, New Delhi.

Ratcliffe, J. 2004, 'Geocoding Crime and a First Estimate of a Minimum Acceptable Hit Rate', *International Journal of Geographical Information Science*, 18, pp. 61–72.

Richardson, H. W. 1988. 'Monocentric vs. Polycentric Models', *Annals of Regional Science*, 57, pp. 165–82.

———. 1980. 'Polarization Reversal in Developing Countries', Papers of the Regional Science Association, 45, pp. 67–85.

——— 1973. *Regional Growth Theory*. London: Macmillan.

Rietveld, P. and J. Boonstra. 1995. 'On the Supply of Network Infrastructure: Highways and Railways in European Regions', *The Annals of Regional Science*, 29, pp. 207–20.

Rivera-Batiz, F. 1988. 'Increasing Returns, Monopolistic Competition, and Agglomeration Economies in Consumption and Production', *Regional Science and Urban Economics*, 18, pp.125–54.

Romer, P. 1986. 'Increasing Returns and Long Run Growth', *Journal of Political Economy*, 98, pp. 71–102.

Rosenstein-Rodan, P. 1943. 'The Problem of Industrialization of Eastern and South-Eastern Europe', *Economic Journal*, 53, pp. 202–11.

Rosenthal, S. S. and W. C. Strange. 2001. The Determinants of Agglomeration', *Journal of Urban Economics*, 50, pp. 191–229.

Roth, P. 1994. 'Missing Data: A Conceptual Review for Applied Psychologists', *Personnel Psychology*, 47, pp. 537–60.
Rubin, D. B. 1987. *Multiple Imputation for Nonresponse in Surveys*. New York: Wiley & Sons.
———. 1978. 'Multiple Imputations in Sample Surveys: A Phenomenological Bayesian Approach to Nonresponse'. Proceedings of the Survey Research Methods Section, American Statistical Association, pp. 20–34.
Saha, S. K. 1987. *Industrialization and Development in Space: An Indian Perspective*. Swansea: Center for Development Studies.
Sala-i-Martin, X. 1996. 'Regional Cohesion: Evidence and Theories of Regional Growth and Convergence', *European Economic Review*, 40, pp. 1325–52.
Santos, M. 1979. *The Shared Space: The Two Circuits of the Urban Economy in Underdeveloped Countries*. London: Methuen.
Schoenberger, E. 1989. 'New Models of Regional Change', in R. Peet and N. Thrift (eds), *New Models in Geography: The Political-Economy Perspective*, London: Unwin Hyman, pp. 115–41.
Scott, A. 1988. 'Flexible Production Systems and Regional Development: The Rise of New Industrial Spaces in North America and Western Europe', *International Journal of Urban and Regional Research*, 12, pp. 171–85.
Seitz, H. and G. Licht. 1992. 'The Impact of the Provision of Public Infrastructures on Regional Development in Germany', Zentrum fur Europaische Wirtschaftsforschung GmbH, Discussion Paper, pp. 93–13, Mannheim Germany.
Secretaria da Receita Federal. 2003. *Demonstrativo das Benefícos Tributários*. Ministeri de Fazenda. Brasilia/DF.
Shankar, R. and A. Shah. 2003. 'Bridging the Economic Divide within Nations: A Scorecard on the Performance of Regional Development Policies in Reducing Regional Income Disparities', *World Development*, 31, pp. 1421–41.
Sharma, T. R. 1954. *Location of Industries in India*. Bombay: Hind Kitabs.
Shukla, V. 1984. 'The Productivity of Indian Cities and Some Implications for Development Policy'. Unpublished Ph.D. thesis, Princeton University.
Storper, M. 1991. *Industrialization, Economic Development and the Regional Question in the Third World: From Import Substitution to Flexible Production*. London: Pion.
Storper, M. and R. Walker. 1989. *The Capitalist Imperative: Territory, Technology, and Industrial Growth*. New York: Blackwell.

Sveikauskas, L. 1975. 'The Productivity of Cities', *Quarterly Journal of Economics*, 89, pp. 393–413.

Tabuchi, 1986. 'Urban Agglomeration, Capital Augmenting Technology, and Labour Market Equilibrium', *Journal of Urban Economics*, 20, pp. 211–28.

Timberlake, M. 1987. 'World-system Theory and the Study of Comparative Urbanization', in M. P. Smith and J. R. Feagin (eds), *The Capitalist City: Global Restructuring and Community Politics*, London: Blackwell, pp. 37–65.

Townroe, P. M. and D. Keen 1984. 'Polarization Reversal in the State of São Paulo, Brazil', *Regional Studies*, 18, pp. 45–54.

Vella, F. 1998. 'Estimating Models with Sample Selection Bias: A Survey', *The Journal of Human Resources*, 33, pp. 127–69.

Venables, A. 1996. 'Equilibrium Locations of Vertically Linked Industries', *International Economic Review*, 49, pp. 341–59.

Verma, S. 1986. 'Urbanization and Productivity in Indian States', in E. S. Mills and C. M. Becker (eds), *Studies in Indian Urban Development*, Washington DC: World Bank, pp. 103–36.

Vernon, R. 1966. 'International Investment and International Trade in the Product Cycle', *Quarterly Journal of Economics*, 80, pp. 180–207.

von Böventer, E. G. 1970. 'Optimal Spatial Structure and Spatial Development', *Kyklos*, 23, pp. 903–24.

von Thunen, J.H. 1966. *The Isolated State*. Oxford: Pergamom Press.

Wade, R. 1990. *Governing the Market: Economic Theory and the Role of Government in East Asian Industrialization*. Princeton: Princeton University Press.

Walker, R. A. 2000. 'The Geography of Production', in E. Sheppard and T. J. Barnes (eds), *A Companion to Economic Geography*, London: Blackwell, pp. 113–32.

Wasylenko, M, T. J. Bartik, H. Duncan, T. McGuire, and R. Ady. 1997. 'Taxation and Economic Development: The State of the Economic Literature', *New England Economic Review*, Mar/Apr, pp. 37–52.

Webber, M. J. 1984. *Industrial Location*. Beverly Hills, CA: Sage Publications.

Weber, A. 1909. *Theory of the Location of Industries.* (translated by Carl Friedrich in 1929), Chicago: University of Chicago Press.

Wei, Y. D. 2000. *Regional Development in China: States, Globalization, and Inequality*. London and New York: Routledge.

Weibull, J. 1976. 'An Axiomatic Approach to the Measurement of Accessibility', *Regional Science and Urban Economics*, 6, pp. 357–79.

Wheaton, W. C. and H. Shishido. 1981. 'Urban Concentration, Agglomeration Economies and the Levels of Economic Development', *Economic Development and Cultural Change*, 30, pp. 17–30.

White, M. J. 1983. 'The Measurement of Spatial Segregation', *American Journal of Sociology*, 88, pp. 1008–19.

Williamson, J. 2000. 'What Should the World Bank Think About the Washington Consensus?', *World Bank Research Observer*, 15, pp. 251–64.

———. 1990. *Latin American Adjustment: How Much has Happened?* Washington DC: Institute of International Economics.

Williamson, J. G. 1965. 'Regional Inequality and the Process of National Development', *Economic Development and Cultural Change*, 13, pp. 3–45.

Wilson, J. D. 1986. 'A Theory of Interregional Tax Competition', *Journal of Urban Economics*, 19, pp. 296–315.

Winship, C. and R. D. Mare. 1992. 'Models for Sample Selection Bias', *Annual Review of Sociology*, 18, 327–50.

World Bank. 2005. 'Brazil Regional Economic Development — (Some) Lessons from Experience'. World Bank Report. August. Processed, Washington DC.

———. 2004a. 'Bihar : Towards a Development Strategy'. World Bank. Processed, Washington DC.

———. 2004b. *India: Investment Climate Assessment 2004 — Improving Manufacturing Competitiveness*. Washington and New Delhi: World Bank.

———. 1999. *Entering the 21st Century: World Development Report 1999/2000*. Washington, DC: The World Bank.

———. 1998. 'Public Expenditures for Poverty Alleviation in Northeast Brazil: Promoting Growth and Improving Services'. World Bank Report 18700-BR. December. Processed, Washington DC.

———. 1987. 'Brazil: Industrial Development Issues of the Northeast'. World Bank Economic and Sector Report. Washington, D.C. Processed, Washington DC.

———. 1986a. 'Korea Spatial Strategy Review'. World Bank report 5868-KO, September. Processed, Washington DC.

———. 1986b. 'Regional Cities Development Project Implemented Between 1985 and 1995'. World Bank Report 4673-TH, March. Processed, Washington DC.

———. 1980. 'The Development of Regional Cities in Thailand'. World Bank Report 2990-TH, June. Processed, Washington DC.

———. 1977. 'Spatial Development in Mexico'. World Bank Report 1081-ME, January. Processed, Washington DC.

Index

accessibility, improved, and effect on reducing geographic barriers 195
advanced regions, benefits to 67, 68
agglomeration, concentration of high income and productivity in 2
agglomeration economies 7, 9
  constituents of 147
  at district level 137
  inter-industry linkages 11–12
  geography and 110, 112
  localization 9–10
  urbanization economies 12–13
agribusiness sector 41
  clustering effects of 66–7, 74n7
Alexander and Company, coalmines of 28
analytic framework, for location of industry 108–11
Annual Survey of India (ASI), data 38–40, 107–8, 118, 140, 148
autarky, conditions under, and location of industry 109–10

Bangalore, information technology in 153
Bihar
  agricultural performance of 211
  as a poor state 33, 67–8
  backwardness in 211
  caste-based governments in 82
  caste conflicts and anarchy in 68
  development outcomes in 187
  investments in 44, 52
  lack of investments in 81
  poverty in 187
  regional development policy options for 211–12
Bombay,
  first cotton mill in 28
  as trade centre 30
  see also Mumbai
Brazil, Constitutional Investment Funds, for social and economic development of lagging regions 200
  industrial growth in 'agglomerative field' in 72
  investment subsidies in 198–201
  Northeast, development in lagging 192
  private sector development in 200
  spatial policy in 200–1
buyer–supplier linkages 112–13, 174–5, 177, 181
  proximity to, and cost reduction 129

calicoes, popularity of, in England 29
Calcutta
  industrial concentration in top districts in 157, 179–80
  decline in industries in 151
  industries clusters in metropolis of 163–6
  investments in suburban districts 55
  jute, iron and steel industries in 151

INDEX

leather industry in 172
polluting industries in fringe areas of 172–3
study of 152–3
as trade centre 30
capital
definition of 118–19
investment 76, 77, 91
private and public 78–80
productivity of 81, 91
Census of India 107
central government investments 69–70, 93, 96
cluster 66, 67–8
in infrastructure 70
location decisions and 104
quality determinants of 98
Central Statistical Organisation (CSO) 38, 148
Centre for Monitoring Indian Economy (CMIE) data 39, 40, 93
chemical industries, pollution from 172–3
chemical and petrochemical industries 41
Chennai 153–4
clustering of metal-electrical combination in 178
clustering of labour in 178
concentration in clusters in 158, 179–80
high location quotients (LQ) for leather sector in 172
industrial centre in 154
industry clusters in metropolis of 155, 167–70
investments in suburban districts in 55, 70
polluting sectors in fringe areas of 173
study of 151
Chinitz–Jacobs diversity 113

clustering,
and access to specialized labour pool 145
benefits of 145
concept of 109
constitution of 147
measurement of 149–51
variable 90
coalmines, opening of 28
coastal districts,
development of 74, 78, 82
bias towards development 69, 71, 92
investments in 54
performance of 54
preferred private sector investments in 103
co-clustering, in selected industry pairs 177
co-location 178, 184
and co-clustering 155, 173–5
Colombia, regional de-concentration of income and industry in 71
colonialism, economic history under 29–30
concentrated decentralization, new growth centres in advanced regions 72, 74n9
'concentration', and 'clustering', of industries 55–7
Constituicao Federal (CFs), in Brazil 200–1
correlation coefficients, for industry pairs 176
in Chennai 175
in Calcutta 175
in Mumbai 175
cost functions, of firms 124–6
cost elasticity of economic geography variables 127
of production factors 126
cost-saving externalities, firm location decisions and 137–8
cotton mills, in Bombay 28

INDEX

'cumulative causation' thesis 5, 70, 91

Davar, Cowasji 28
decentralization, concentrated 71–3
Delhi, investments in suburban districts in 55, 70
developing countries, low levels of investments in 2
   spatially explicit policies for lagging regions 192
district level incentive system 102
district population, role of, for new industry 91
diverse economic regions, cost reducing effects of 129

East India Trading Company 29
economic activity, clustering of 205–6
   spatial concentration of 6–7
economic diversity, of region 113–14
   and firm location 109
economic geography of India 3, 30
   and demand for traditional inputs 130–2
   estimates for 124, 125, 127
   factors 36, 137, 139
   and firms 106
   variables 108, 111–18, 133
economic growth, industrialization and 3
econometric analysis, results from 124–5, 128–30, 133
economies of scale, at metropolitan area level 145–6, 197
economy 4
   'opening' of 35
efficiency related factors, and investment decisions 103
Ellison-Glaeser (EG) index of concentration 122–3
employee output, patterns among industries 122
employment, in industries 31, 33

energy, costs 119, 133
   requirements and economic geographic variables 133
English investments, in India 30
European Union, impact of infrastructure investments 194
exchange rate policy 207
'explicit' spatial policies 5
export, of cotton to China and Far East 28
Export Processing Zone (EPZs), investments in 37
expropriation, history of 209
   institutions of 210

Factories Act of 1948, amendment to 38
factory/firm/industry, clustering of 155, 172
   economic geography of 106
   level output 118
   location 137, 182
     decisions, determinants of 78–9, 208–9
     and transportation cost 79
     and worker clusters, relative location of 184
financial incentives 192
   for location in lagging districts 204
Five-Year Plan, Fourth, on regional development in India 208
fixed capital, in industries 31
'folk theorem', of spatial economies 7
food/beverages sector, clustering of workers and factories and clustered pin codes 171–2
foreign direct investment (FDI) 66, 75, 93
   favouring coastal, metropolitan-oriented areas 54, 69
foreign exchange reserves, shortage of 34

free trade areas 37
Freight Equalization Policy, 1956 34, 207
  amendment to 35

Gandhi, Rajiv 44, 72
  liberalization under 35
geographical cluster, of industries 91, 122–3
geography 76, 78
  role of, in investment location 102
  unevenness, and investment 75
Gini Coefficient, to measure industry concentration 56
global clustering 155–6, 178
  measures of 155
globalization, developing countries' response to 5
government investments 54
Greater Bombay,
  decline in investment share of 69
  investments in 52, 102
  new investments in 67
gross domestic product (GDP), growth of 4
Gujarat, entrepreneurial culture in 103
  investments in 44

heavy industries 41
  clustering of 58–9, 66
  investments and gains in 53, 73n3
Heckman selection model, of new investments 84–5, 88–90, 92
Herfindahl index, to measure economic diversity 114, 123
'Hindu rate of growth' 36
'home market effect' 8

'implicit' spatial policies 5
income, disparities in 4, 206
  indicator in India 31–2
India, balanced regional development policy in 206–7
  inter-regional polarization in 72
  regional income disparities in 206
  spatial policies for regional development in 203–5
Indonesia,
  'bureaucratic capitalism' in 206
  dominance of Java in 205–6
  firm-level choice in 196–7
  firm migration to lagging regions in 197
  industrial de-concentration in 71–2
industrial city 3
industrial districts 173–5, 183
  buyer–supplier linked 175
  labour sharing 175
Industrial Policy Resolution of 1956 34
industrial sectors, concentration of 120–1
  model findings for 90–3
industries/ industrial, classification code 41
  co-location 180
  concentration and agglomeration 55, 123
  credit 81, 91
  data on 38–41
  decentralization of 33
  de-concentration of 30
  diversity, and cost saving for firms 134
  groups, co-location of 175–6
  growth, in lagging regions 203
  labour force 77, 91, 101
  local level linkages of 113
  location, decision 31, 72–3, 75, 78, 101–4, 106–7, 146–7, 188, 210
    model of 80–9
  spatial regulations on 185
    and transport improvements 196
    in urban fringes 180
  output, growth of 4
  policy 187

and reforms 33–6
spatial variations in 122–4
see also investments
industry clusters, and 'concentration' 134, 144–5, 149, 177
in Calcutta Metropolis 163–6
in Chennai Metropolis 167–70
in Greater Bombay 159–62
influences for 205
local 155, 156, 171–3
in metropolitan regions 142
success of 109
industrialization,
in India 3, 28–33
inequalities 3
infant mortality, and investments 82, 99–101
infrastructure pattern,
in India 31–2, 76–7
aggregate benefits from 195
investments in 99–101, 191–2, 193–8
impact of 198
and regional development 208
physical, and attraction of new investments 91–2
role of 99–100
value of 81
variables for new industry 91
inland regions, decline of 53
input,
demand substitution 131–2
output linkages, and proximity to buyers and sellers 125, 129
inter-industry, linkages 112–13
transfers 107
inter-regional divergence 73
inter-related firms 110
and profit maximization 108–9, 112
intra-industry transfers 107
intra-metropolitan industrial location decision 185
intra-regional divergence 73

investments, in industries 28, 37, 41, 44–5, 52, 57–60
in coastal districts 93
in infrastructure, and development in lagging regions 193–4
new 85, 93
non-zero, in districts 84–5
in post-reform period 63–4, 66
in pre-reform period 62–4
private sector 102–4
sectoral disaggregation of post-reform 54
spatial patterns of 70–1
state-wise distribution of 46–8
Italy, transport infrastructure and regional inequalities in 195–6

Japan, polices for development of lagging regions in 199
joint sector 44
investments in 45, 54

Kerala, communist government and anti-capitalism policy of 68
literacy effect and low investments in 92
poor performance/ lagging by 67–8
knowledge spillovers, local co-clustering of factories and 145, 173–5
information spillovers 174
technology spillovers 174
Kolkata *see* Calcutta
Korea, Local Share Tax Law in 203
policies for balanced regional growth in 203
revenue sharing in 198–9, 203
Krugman's thesis 8

labour/workers, characteristics, and investments 100–1
clustering of 155, 171–2
co-location of 175
concentration 180

definition of 118
in manufacturing industry 81
pools 181, 183
productivity of 81, 91, 120, 123–4
regulations 209
requirements, economic geography variable and 130, 133
skilled and unskilled 174
variables 91, 99
labour markets, co-clustering of factories and 173–4
lagging regions, firm migration to, and improvements in 196–8
improving conditions of 188, 190–1
state development of 71, 192–3
land, market constraints, and location of firms 147, 184, 189
tenure under British rule, and economic performance 210
-use policies 142–4, 184–5
leather industry, location in urban fringes 184
pollution from 172
liberalization 35
literacy, effect 92
investments in 82, 101
living standards, disparities in 2
local business regulations, and firm location 209
local economic diversity 125, 129–30
local industrial base, diversity of 107
localization economies 143–5, 173, 181
and clustering 142, 144–7
labour markets and 174
and urban economies 181–2
versus urbanization economies in cluster formation 146–7
logistic models 99–100
of new investments 86–7

macroeconomic policies, impact of 207
Madhya Pradesh, public sector development in 204

Madras, as trading centre
see also Chennai
Maharashtra 30
development of growth centres in 204
investments in 44
new investments in 67
Malaysia, development to reduce isolation of northeast peninsula 192
development of secondary cities in 192
manufacturing industries, analytic framework for locating 108–11
location choices of 107–41
role of 3
mapping methods 41–3
markets,
access to 4, 76, 107, 110, 114–18, 135
impact of 125, 128, 195
and transport costs 7–9
forces 3, 4
and states 3–6
metals and electrical/electronic industries, input and output linkages in 178
metals and machinery industries, input and output linkages in 178
metropolitan regions 70, 74n5, 82–3
clustering of industries in 142, 180
decisions on location of industries in 182
discouragement of heavy industries in 203–4
favouring investments in 54–5, 68–9
investments in 93–4
location of industries in 35, 78, 122, 189
performance of 52
planning 212
and urban agglomeration 122
workers and factories clusters in 180

Mexico, fiscal incentives to move industries outside urban centres 201–2
  free trade and investment with NAFTA 201
  moving industry away from agglomeration in 193
  tax and import duty exemptions in 198–9, 201–2
migration 212
minimum needs programmes 208
Ministry of Urban Affairs and Employment 35
mixed industrial districts 134, 182
Monopolies and Restrictive Practices Act, amendment to 35
Moran's I 42, 83, 156, 171
  Global 149–50
  Local 150
Mumbai, concentration of industries in top districts in 178–9
  high location quotients for industries 172
  industry clusters in Greater Bombay 157, 159–62
  investments in 55, 70, 151
  metals–electrical factories, clustering of 178
  polluting industries in fringe areas of 172–3
  population of 151
  study of 151–2
  textile clusters in 184

National Commission on Urbanization (1988) 34
National Industrial Classification (NIC) code 40
national input–output accounts 112–13
neighbourhood, definition of 150
North America Free Trade Agreement (NAFTA), between Mexico and North America 201

OLS selection model, of new investments 85–5, 88–90, 92
Orissa, development of public sector in 204
  heavy industry sector in 67
  investments in 52, 67
own-industry, concentration of 110–11, 125, 128–9
  employment and localization economies 111
  and profit maximization 108–9

pecuniary externalities 108, 130, 135n2
planning, in India 33
Planning Commission 33, 208
plant level data 140–1
'polarization reversal', inter-regional polarization and intra-regional dispersal of industries 72, 74n8
policy, environment 189–92
  instruments 190–1
polluting industries, location of 172–3, 184
port, access to, and firm location 196
printing industries, clustering of workers and factories in 171–2
private sector investment 34, 45, 54, 66, 81, 93–101, 146
  favouring south, southeast and coastal areas 69
production externalities, and firm location 205–6
productivity, indicators 122
  spatial concentration of and higher 145
public sector investments 93–101
Punjab, investments in 52

Raigarh, preferred investments in 102
reforms of 1991 31
  region(al), attractiveness, for firm location 196–7
  development activities, on spatially targeted programmes 189

disparities, policies for reducing 192–203
economic geography, impact on cost structure of firms 130
incentives, for firm location to lagging/underdeveloped regions 191, 198–203
policy instruments 191
regional inequality 31–4
  decline in 33, 187–8
  and state intervention 5, 190
regionally targeted interventions, limited pay-off from 204–5
Reserve Bank of India, Mumbai 151
road segment, and market access 117

Santhals,
  illiteracy and poverty among 1
  as Scheduled Tribes 1
secondary cities, growth potential of 2
Shephard's lemma 138
'sick' industry 35
'single window' project clearance, at state levels 35
skilled and unskilled labour 174
small-scale operations, clustering of 171
socialism 82
socialist variables, and new investments 92, 101, 105n
South Korea, liberalization and decentralization of manufacturing industry in 71
Spain, impact of investments in infrastructure and regional convergence in 196
  public capital, positive effect of 194
spatial autocorrelation 42, 43, 83–4, 149
spatial clustering, extent of 55–60
spatial concentration, data and statistics 108, 118–24
  and transportation 110
'spatial Gini' 149
'spatial lag' 83–4, 99, 100

'spatial' model, for government investments 99
  for new investments 94
spatial policies 187–212
  limitations and returns from 203–5
spatially targeted tax abetment policy 207
state capitalism 75
subcontracting 129, 145
subsidies, for lagging regions 192
supplier and demand linkages, for location of manufacturing industries 108–9

Tamil Nadu 30
tax holidays 192–3
  for development of lagging/underdeveloped regions 198–202
technology parks 37
textile sector, concentration and clustering of 58, 66–7
  in Mumbai CBD 184
  and separate cluster for workers 181
Thai Board of Investment, Thailand 202
Thailand, development of secondary cities in 192
  fiscal incentives to regions outside Bangkok 193, 202
  tax holidays in 198–9, 202
Thane, preferred investments in 102
total labour pool 77
trading linkages 205
transport costs, and firm location 7–9, 108–10
transport improvements, and firm relocation 208
transport infrastructure, digital 117
  firm location and, in Indonesia 196–7
  and industry clustering 205
  and market integration 116

and reduced regional inequalities 195–6
trans-shipment hubs, distances to 117–18
and urban centres 118
uncoordinated policies 208–9
United States, impact of infrastructure investments on productivity and growth 194
manufacturing industry, concentration pattern of 123
universal suffrage, impact on economic growth 210
urban agglomeration 206
urban centres, travel time to 117
urban development 35
urban economies, and localization economies 181–2
urban land-use 78
urbanization 145
economies 113, 143, 189, 206
Uttar Pradesh, caste conflicts and anarchy in 68
poor performance/lagging by 67–8

wages, of labour in metropolitan areas 122
Washington Consensus discourse 37
West Bengal 30
coalition communist government's policy in 68, 152
decline in prosperity and income in 33
investments in 44, 52
poor performance/ lagging by 67–8

zamindari system, and revenue collection under British rule 210

MAR 1 2 2009